ONE OF THE **GUINNESS** FAMILY OF BOOKS

AIR
Facts & Feats

ONE OF THE GUINNESS FAMILY OF BOOKS

AIR
Facts & Feats

JOHN W. R. TAYLOR
MICHAEL J. H. TAYLOR
DAVID MONDEY

STERLING PUBLISHING CO., INC.   NEW YORK

OTHER BOOKS OF INTEREST

COVER PHOTOGRAPHS

Front cover (clockwise from top left)
Air France Concorde
Bensen Gyro-Copter autogyro in flight (Howard Levy)
Sikorsky Bolshoi, first four-engined plane to fly
Guppy-201, with the largest cargo compartment of any aircraft
Santos-Dumont No. 6, first airship to round the Eiffel Tower

Back cover (clockwise from bottom right)
Lockheed Q-Star, so quiet it was not heard 100 ft overhead
Northrop T-38 Talons of the USAF aerobatic team,
  the Thunderbirds
Supermarine Spitfire MkXVI that remains in flying condition
  today (Howard Levy)
Rotorway Scorpion, a "build-it-yourself" helicopter
Vickers Vimy taking off from Newfoundland

Revised Edition © 1978 by Sterling Publishing Co., Inc.
Two Park Avenue, New York, N.Y. 10016
First published in Great Britain under the title "The Guinness
Book of Air Facts and Feats" © 1977, 1973, 1970 by John W. R.
Taylor, Michael J.H. Taylor, David Mondey and Guinness
Superlatives Ltd.
Manufactured in the United States of America
All rights reserved
Library of Congress Catalog Card No.: 77-93306
Sterling ISBN 0-8069-0126-8 Trade
0127-6 Library

CONTENTS

INTRODUCTION

In the four years since the last edition of *Air Facts and Feats* was published, a number of significant projects have proved, once again, that there is still an enormous reserve of technological acumen and ability within the world's aerospace industry. That this is so can often be measured in the accountant's clinical financial terms of profit rather than loss. This is, however, an indifferent criterion by which to judge man's achievements in this field. More important, by far, is the contribution that aerospace has made towards the prevention of a third world war, the exploration and development of natural resources, the saving of lives, and the provision of safe, fast and comfortable travel on a worldwide basis.

More than a century ago the British historian Lord Macaulay commented that: '. . . those inventions which abridge distance have done most for the civilisation of our species . . . (tending) . . . to remove national and provincial antipathies, and to bring together all the branches of the great human family.' Aerospace has done much to speed this continuing process towards a more universal awareness that peoples and nations must live and work together.

Working together has, in fact, become a way of life for those involved in European aerospace, where the research and development cost of new projects has proved too great for any single nation to bear. From such co-operation three important aircraft have appeared on the European aviation scene during the last four years: the Airbus Industrie A300B, Concorde SST and Panavia Tornado. The Airbus, an important short/medium-range wide-body transport, represents collaboration between France, Germany, the Netherlands, Spain, and the United Kingdom. Concorde, both exciting and practical, which has in a single step virtually doubled the speed of air travel for civilian passengers, is the work of France and the United Kingdom. The Panavia Tornado multi-role combat aircraft represents Europe's biggest industrial programme, and has combined the competence of Germany, Italy and the United Kingdom to produce an advanced military aircraft of which some 800 examples are expected to enter service. All of these projects have benefitted from the involvement of multi-national brains and expertise.

The aircraft industry of the United States has notched up important new milestones without recourse to international programmes. We have seen in these four years the emergence of the F-16A air combat fighter from General Dynamics, an advanced technology military aircraft of which more than 3000 are likely to be built. The Boeing Company and McDonnell Douglas are involved competitively in the development of short take-off and landing (STOL) military transport prototypes which hint at exciting prospects for future generations of civil transport aircraft. Rockwell International has rolled out the first Space Shuttle Orbiter for America's National Aeronautics and Space Administration (NASA), the first re-usable space transportation system in the world. It had already flown in free (but unpowered) flight when this book closed for press.

In the pages which follow you can read much, much more about the thrilling events that have made a highway of the skies – evolving from the tiny seeds of man's longing to fly, germinated so many generations ago.

Compared with its predecessors, this third edition of *Air Facts and Feats* offers a far wider coverage of the whole aviation scene, with increased emphasis on pioneering flights and the research machines which pointed the way to great and more familiar production aircraft which won wars and made possible an airline industry which now carries around the world annually more than 500 million passengers. All but a handful of the illustrations are new, and the entire text has been reviewed for accuracy, rearranged in a more logical sequence, and supplemented with recently-discovered or up-to-the-minute facts and figures.

In the distant years ahead, when our lives and time have become past history, recorded for study by generations yet unborn, they who read and think will look back on this 20th century as one of the most breathtaking in the whole history of man. The century in which man learned to fly; to evolve electronics miracles with which to communicate, calculate and control; to put men on the Moon and to explore the planets.

Bladud, fabled ninth king of Britain, was an early 'birdman'. Wearing home-made wings, he jumped from the highest point on the temple of Apollo in Trinaventum and was killed

SECTION 1

MYTH AND FANTASY

Myths and fantasies might seem to have no place in a book of facts and feats; but, as the lyric-writer of a modern American musical reminded us, if you do not have a dream you cannot have a dream come true. The dream of flying must date back almost to the day when a prehistoric ancestor of man saw a bird for the first time. How convenient it would have been to take wing the next time he encountered a river he could not cross, came to a sea shore and could only guess at what lay beyond the water, or, most frustrating of all, found himself trapped by a wild animal, or enemies, with no way of escape.

Little wonder that when he devised gods to worship, they usually dwelt somewhere 'up there', above the trivialities and pain of the everyday world. Gods like the Greek Hermes, whose duties required occasional descent from Olympus, were pictured quite naturally with wings – attached to their persons or to their accoutrements – so that they could travel by the easiest method imaginable. Even when the gods of mythology were rejected by the people of biblical Old Testament days, the dream was not buried with them. The psalmist could only sing of his longing for the wings of a dove. Two centuries later, or thereabouts, while the prophet Elijah was walking with his chosen successor Elisha, a chariot of fire pulled by horses of fire came between them, and Elijah was taken up to heaven by a whirlwind.

Whether or not one accepts that story as being true, it is interesting to note that Elijah did not sprout wings, or strap on a pair of artificial wings in order to fly, but was provided with a flying machine which some modern romantics might like to associate with flying saucers. The Persians tell of one of their ancient kings, Kai Kawus, whose favourite relaxation seems to have been to travel by air around his kingdom on a throne lifted and propelled by four geese. Lacking such powerful domestic birds, or perhaps unaware of the exploits of his oriental predecessor, the fabled ninth king of Britain was simply one of a long line of 'birdmen' who tried and failed.

His name was Bladud, and his reputed coronation year of 863 BC makes him a contemporary of Elijah. Nearly 2500 years later, a poem published in London let Bladud seem to tell the story of his flying activities in his own words:

> I deckt my corps with plumes (I say) and wings,
> And had them set, thou seest, in skilfull wise
> With many feats, fine poyseing equall things,
> To aide my selfe in flight to fall or rise,
> Few men did ever use like enterprise,
> Gainst store of wind, by practise rise I could,
> And turne and winde at last which way I would.

> But ere the perfect skill I learned had,
> (And yet me thought I could do passing well)
> My subjects hearts with pleasant toyes to glad,
> From Temples top, where did Apollo dwell,
> I sayd to flie, but on the Church I fell,
> And in the fall I lost my life withal.
> This was my race, this was my fatall fall.

In more terse, modern language, the King put on his wings, jumped from the top of the Temple of Apollo to amuse his subjects, and made a fatal crash-landing on the roof. As the poet warned:

> On high the tempests have much powre to wrecke:
> Then best to bide beneath, and surest for the necke.

However, the would-be 'birdmen' refused to be deterred, and continued jumping from every kind of high place, flapping their wings desperately but vainly all the way to the ground. And why not, when they could read stories of how a Greek named Daedalus had built wings for himself, and his son Icarus, and flapped to freedom from a prison on the island of Crete. On that occasion the safety record was only 50 per cent. Being adventurous, young Icarus had climbed too high, so that the sun melted the wax holding the feathers in place. He was, in fact, the first recorded victim of a structural failure in flight; but by no means the last among the birdmen. Countless numbers died before an Italian named Giovanni Borelli published a book named *De Motu Animalium*, in 1680, in which he explained at length why man could never hope to sustain his own weight in the air without mechanical assistance.

What was the alternative?

At this stage, the story begins to progress from myth to a fascinating blend of fact and fantasy, with the Middle East – cradle of so much mystery – as the starting point.

Rummaging among boxes of small artefacts in a store-room at Cairo Museum in 1972, Dr Khalil Messiha came across a strange-looking wooden bird model that had been found originally at Saqqara in 1898. Unlike a real bird, or any of the other models carved in the 3rd or 4th century BC, this one had a rear body of narrow elliptical shape, like that of a small modern aeroplane, and ended in a deep vertical tail-fin containing a groove for a horizontal tailplane. No bird can contort its body to such a shape.

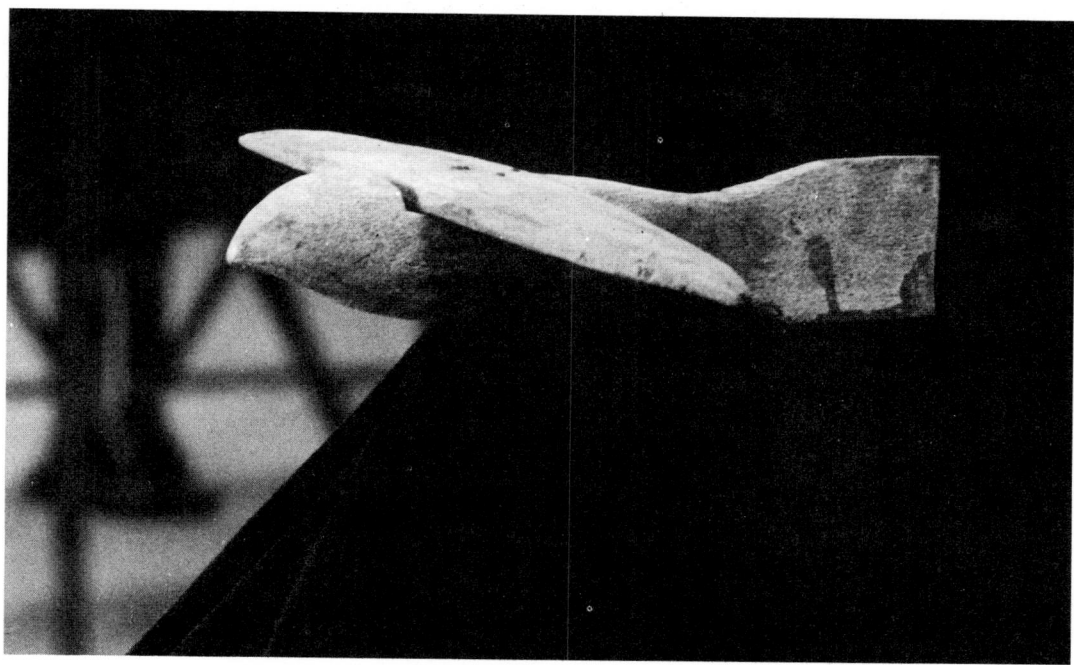

Wooden bird, possibly a glider, carved in Egypt some 2300 years ago (Boudewijn Weehuizen)

Did this chance discovery suggest that the ancient Egyptians built model gliders – perhaps even full-scale gliders? Some sceptics prefer to regard the model as no more than a wind direction indicator, pivoted in such a way that the vertical tail would turn it into wind. Why then, queried Dr Messiha, had the ancient Egyptian craftsman spent so long giving the wing a finely curved aerofoil section?

Whatever we choose to believe about that model of 2300 years ago, it is certainly fact. It spans 7 in (18 cm), weighs under 1½ oz (40 g) and is made of sycamore. Amazingly, we have to pass over some 1600 years of history before we find positive proof of anyone else thinking in terms of machines in which men might fly. In about the year 1250 AD, an English Franciscan monk named Roger Bacon wrote a book entitled *Secrets of Art and Nature*, in which he mentioned 'Engines for flying, a man sitting in the midst thereof, by turning only about an Instrument, which moves artificiall Wings made to beat the Aire, much after the fashion of a Bird's flight.' Bacon insisted that he knew personally the 'very prudent man' who had invented this flapping-wing aircraft, or ornithopter, although he admitted that he had not seen it himself. At the very least Bacon was, so far as we know, **the first person to write about flying in any mechanical or scientific sense.** His whole purpose, he claimed, was to prove, in an age of alchemists, the 'inferiority and indignity of magical power compared with nature and science.

Even with our 20th century scientific knowledge it has not yet proved possible to build an ornithopter capable of flight with a man on board. Bacon could not know the impracticability of the flying machine he described; neither could the great Italian artist-inventor Leonardo da Vinci (1452–1519), who produced many designs for flapping-wing aircraft. At least he did his best to supplement ambition with science. One of Leonardo's carefully drawn designs shows the wings mounted on a wooden structure fitted with levers and pulleys through which the would-be pilot's muscle-power might be used more effectively. Legs supplemented arms in flapping the wings, and the pilot could even steer up or down by nodding his head. Leonardo's true cleverness is indicated best, perhaps, by the fact that he never tried to fly in such a contraption.

By the middle of the 17th century, the dream of flying like the birds was still so far from fulfilment that any alternative to flapping wings would have been welcome. Von Guericke's invention of the air pump, by which a partial vacuum could be created inside a sealed vessel, would hardly have offered an alternative to most people; but, in the year 1670, it gave the Jesuit priest Francesco de Lana-Terzi an idea. . . .

This entailed lifting a boat-shape carriage into the air by means of four large copper globes from which all the air had been extracted by air pump. De Lana believed that if the copper was thin enough the globes would be considerably lighter than the air they displaced and would, therefore, float. Unfortunately, it proved impossible to make globes of sufficient size and lightness that would also be strong enough to withstand atmospheric pressure trying to crush them. At which point de Lana announced that he did not want to build flying machines anyway, in case they were used to drop boiling oil and other unpleasant loads on to ships or armies in time of war.

De Lana's flying boat was the **first definite design for a lighter-than-air craft.** Although itself impractical, it pioneered a line of development that was to turn the age-old dream of flight into reality in the century that followed.

Icarus, with his father, Daedalus, who built his wings. Both attempted to flap their way to freedom from imprisonment by Minos, King of Crete, in mythological times. Daedalus reached safety in Greece. Icarus, climbing too high, suffered a structural failure and crashed into the sea when the sun melted the wax holding his wings together.

Cameron D-96, the world's first hot-air airship (P J Bish)

SECTION 2

LIGHTER THAN AIR

In all mankind's modern preoccupation with wing-borne and rocket flight, one tends to forget that 'exploration' of the air had begun more than one hundred years before the Wright brothers achieved manned, powered flight in their aeroplane; that the advance of science towards the end of the 18th century had enabled men (and women) to be carried aloft and float fairly peacefully across country – albeit at the mercy of the elements. For some years after the advent of Montgolfier hot-air balloons in Paris, opinions were divided on the relative merits of heated air and of hydrogen as the best lifting agent, but inevitably the dangers inherent in carrying a bonfire suspended under the envelope decided early pioneers to resort to the more expensive process of chemical reaction for the generation of hydrogen. It was not until well into the 20th century that the inert helium gas gained preference over hydrogen (use of which had caused countless aerial tragedies down the years), but it was then too late to preserve lighter-than-air travel in the age of the passenger aeroplane.

Ballooning during the 19th century was undertaken on a far greater scale than is realised today.

Aeronauts were not merely the daring sportsmen of their age; they were frequently called upon to demonstrate the practical uses to which their craft could be put. Balloons were used in wartime – even on the battlefield – and for the carriage of mail and military despatches. The English Channel was crossed by air 124 years before Louis Blériot made his perilous flight.

In describing the achievements of those early pioneers it is necessary to remember that the almost total lack of scientific education among 'civilised' populations allowed an almost medieval fear of the sky to persist; and a man who allowed himself to be carried aloft (as if by magic) would certainly face unknown terrors. Moreover, the unheralded arrival of a balloon from the sky might engender widespread hysteria among the local populace. One can learn of similar reaction to the aeroplane among primitive tribes in the modern age.

It was the helplessness of early balloonists, drifting at the whim of the wind, that determined the more constructive and adventurous to seek propulsion and directional control. Their success led to the airship or 'dirigible', which was at one time regarded as the most destructive vehicle of war as well as the ultimate in safety and comfort in air travel.

Today the balloon and the airship are usually considered to be the playthings of a diminutive band of diehards, swept aside as the chaff of jet and rocket blast. Yet, hot-air ballooning has never attracted more participants than it does today, and small airships are seen again in the skies over many countries.

First successful demonstration of a model balloon took place on 8 August 1709 in the Ambassadors' drawing-room at the Casa da India, Lisbon, in the presence of King John V of Portugal, Queen Maria Anna, the Papal Nuncio, Cardinal Conti (later Pope Innocent III), princes of the Court, members of the Diplomatic Corps, noblemen and courtiers. The balloon, made and demonstrated by Father Bartolomeu de Gusmão, consisted of a small envelope of thick paper inflated with hot air produced by 'fire material contained in an earthen bowl encrusted in a waxed-wood tray' which was suspended underneath. The balloon is said to have risen quickly to a height of 12 ft (3·5 m) before being destroyed by two valets who feared that it might set the curtains alight. Suggestions that Gusmão became airborne later in a full-scale version of his balloon, although documented, cannot be substantiated. The **Passarola** (Great bird), sometimes illustrated as a peculiar form of flying machine devised by Gusmão, is no more than a fanciful representation of the carriage designed for the full-size balloon.

Hydrogen was first isolated in 1766 by the English scientist Henry Cavendish who referred to it as 'inflammable air' or Phlogiston. It was first named hydrogen by the French chemist Lavoisier in 1790.

The first balloon to leave the ground capable of sustaining a weight equivalent to that of a man was a hot-air balloon made by the brothers Joseph and Étienne Montgolfier (1740–1810 and 1745–99 respectively). This balloon, calculated as being able to lift 450 lb (205 kg), was released on 25 April 1783, probably at Annonay, France, rose to about 1000 ft (305 m) and landed about 1000 yd (915 m) from the point of lift-off. The balloon had a diameter of about 39 ft (12 m) and achieved its lift using hot air provided by combustion of solid waste (probably paper, straw and wood) below the neck of the envelope.

The first public demonstration by the Montgolfier brothers took place at Annonay on 4 June 1783, when a small balloon of about 36 ft (11 m) diameter, made from linen and paper, rose to a height of 6000 ft (1830 m).

The first free ascent of a hydrogen-filled balloon (unmanned) was made on 27 August 1783 from the Champ-de-Mars, Paris when Jacques Alexandre César Charles (1746–1823) launched a 12 ft (3·5 m) balloon. It was filled with hydrogen that Charles had manufactured and was capable of lifting 20 lb (9 kg). The balloon drifted for 45 min and came to earth at Gonesse, 15 miles (25 km) from Paris, where it was promptly attacked by a frenzied mob of panic-stricken peasants.

The first living creatures to become airborne under a balloon were a sheep, a duck and a cock which were lifted by a 41 ft (13 m) diameter hot-air Montgolfier balloon at the Court of Versailles on 19 September 1783 before King Louis XVI, Marie-Antoinette and their Court. The balloon achieved an altitude of 1700 ft (520 m) before descending in the Forest of Vaucresson 8 min later, having travelled about 2 miles (3 km). The occupants were scarcely affected by their flight nor by their landing.

The first man carried aloft in a balloon was François Pilâtre de Rozier (born 30 March 1757, died 15 June 1785) who on 15 October 1783 ascended in a tethered 49 ft (15 m) diameter Montgolfier hot-air balloon to 84 ft (26 m) – the limit of the restraining rope. The hot air was provided by a straw-fed fire below the fabric envelope and the balloon stayed up for nearly $4\frac{1}{2}$ min.

The first men carried in free flight by a balloon were de Rozier (see above) and the Marquis d'Arlandes who rose in the 49 ft (15 m) diameter Montgolfier balloon at 13.54 h on 21 November 1783 from the gardens of the Château La Muette in the Bois de Boulogne. These first aeronauts were airborne for 25 min and landed on the Butte-aux-Cailles, about $5\frac{1}{2}$ miles (8·5 km) from their point of departure, having drifted to and fro across Paris. Their maximum altitude is unlikely to have been above 1500 ft (450 m).

The first men to be carried in free flight by a hydrogen-filled balloon were Jacques Charles and one of the Robert brothers who had helped to make it. They ascended from the gardens of the Tuileries, Paris, at 13.45 h on 1 December 1783, in a balloon 27 ft 6 in (8·6 m) in diameter before a crowd estimated at 400 000. The craft landed 27 miles (43 km) distant, near the town of Nesles.

The first women to ascend in a balloon (tethered) were the Marchioness de Montalembert, the Countess de Montalembert, the Countess de Podenas and Mademoiselle de Lagarde who were lifted into the air by a Montgolfier hot-air balloon on 20 May 1784 from the Faubourg-Saint-Antoine, Paris.

The first woman to be carried in free flight in a balloon was Madame Thible who ascended in a Montgolfier with a Monsieur Fleurant on 4 June 1784 from Lyon, France. The balloon, named *Le Gustav*, reached an altitude of 8 500 ft (2600 m) and was watched by the King of Sweden.

The first balloon ascent in Italy was made on 25 February 1784 by a Montgolfier carrying Chevalier Paul Andreani and the brothers Augustin and Charles Gerli at Moncuco, near Milan, Italy.

The first British aeronaut is claimed to have been James Tytler, a Scotsman, who on 25 August 1784 made a short ascent in a Mont-

19 September 1783. A Montgolfier hot-air balloon, carrying a sheep, a duck and a cock in a basket, departing from Versailles. This was the first time that living creatures left the ground in an aircraft

golfier-type balloon, probably from Heriot's Garden, Edinburgh. The maximum altitude reached is believed not to have exceeded 500 ft (150 m).

The first aerial voyage by a hydrogen balloon over Great Britain was that of Vincenzo Lunardi, an employee of the Italian Embassy in London, who on 15 September 1784 ascended in a Charlière from the Honourable Artillery Company's training-ground at Moorfields, London. His flight was northwards to the parish of North Mimms (today the site of the village of Welhamgreen), Hertfordshire. Here Lunardi landed his cat and jettisoned his ballast; this caused him to ascend again and he finally landed at Standon Green End near Ware, Hertfordshire. On the spot where he landed stands a rough stone monument on which a tablet proclaims:

Let Posterity know
And knowing be astonished!
That
On the 15th day of September, 1784
Vincent Lunardi
of
Lucca in Tuscany
The First Aerial Traveller in Britain
Mounting from the Artillery Ground
in London
And traversing the Regions of the Air
For two Hours and fifteen Minutes

in this Spot
Revisited the Earth.
On this rude Monument
For ages be recorded
That wonderous enterprize, successfully
achieved
By the powers of Chymistry
And the fortitude of man
That improvement in Science
Which
The Great Author of all Knowledge
Patronising by his Providence
The Inventions of Mankind
Hath generously permitted
To their benefit
And
His own Eternal Glory

The combined hot-air/hydrogen balloon of
de Rozier and Romain, 15 June 1785

The first English aeronaut was James Sadler who, on 4 October 1784, flew in a Montgolfier-type balloon of 170 ft (52 m) circumference at Oxford.

The first application of a propeller to a full-size man-carrying aircraft was recorded on 16 October 1784, when Jean-Pierre Blanchard added a small hand-operated six-blade propeller to the passenger basket of his balloon. As a means of propulsion it was, of course, completely ineffectual.

The first aerial crossing of the English Channel was achieved by the Frenchman, Jean-Pierre Blanchard, accompanied by the American, Dr John Jeffries. On 7 January 1785 they rose from Dover at 13.00 h and landed in the Forêt de Felmores, France, at approximately 15.30 h, having discarded almost all their clothes to lighten the craft *en route*. Their balloon was hydrogen-filled.

The first British woman to travel under a balloon in Britain was Mrs Letitia Ann Sage who ascended in Lunardi's hydrogen balloon from St George's Fields, London, on 29 June 1785. Lunardi, who had proclaimed that he would be accompanied by three passengers (Mrs Sage, a Colonel Hastings and George Biggin), discovered that his balloon's lifting power was not equal to the task. Rather than draw attention to the lady's weight (by her own admission, she weighed more than 200 lb (90 kg)), he stepped down from the basket with Colonel Hastings. The balloon eventually came to earth near Harrow, Middlesex, where the two occupants were rescued from an irate farmer by the boys from that famous school.

The first aeronaut to be killed while ballooning was François Pilâtre de Rozier who was killed while attempting to fly the English Channel from Boulogne on 15 June 1785 in a composite hot-air/hydrogen balloon. It is believed that, when hydrogen was vented from the envelope, the escaping gas was ignited, and the balloon fell at Huitmile Warren, near Boulogne. Also killed was Jules Romain, Pilâtre's companion.

The first free flight by a balloon in the United States of America was made on 9 January 1793 by the Frenchman, Jean-Pierre Blanchard, who ascended in a hydrogen balloon from the yard of the old Walnut Street Prison, Philadelphia, and landed in Gloucester County, New Jersey, after a flight of 46 min. Among the vast crowd who turned out to witness this event were the President, George Washington, and four future presidents of the United States: John Adams, Thomas Jefferson, James Madison and James Monroe.

The first military use of a man-carrying balloon for aerial reconnaissance was by the French Republican Army at Maubeuge, Belgium, in June 1794. On 26 June, during the battle of Fleurus, Captain Coutelle ascended in the observation balloon *Entreprenant*.

The first woman to lose her life in an aerial disaster was Madame Blanchard, widow of the pioneer French aeronaut Jean-Pierre Blanchard (who had died after a heart attack, suffered while ballooning, on 7 March 1809). Madame Blanchard was killed when her hydrogen balloon was ignited during a firework display which she was giving at the Tivoli Gardens, Paris, on 7 July 1819.

The first long-distance voyage by air from England was made during 7–8 November 1836 by a hydrogen balloon – *The Royal Vauxhall Balloon* – manned by Charles Green (English aeronaut) accompanied by Robert Holland, MP, and Monck Mason, who ascended from Vauxhall Gardens, London, and travelled 480 miles (770 km) to land near Weilberg in the Duchy of Nassau. The balloon was subsequently named the *Great Balloon of Nassau.*

The first balloon bombing raid was carried out on 22 August 1849 when Austrian hot-air balloons (pilotless), each carrying a 30 lb (14 kg) bomb and time-fuse, were launched against Venice. They caused little damage.

The first long-distance balloon flight in America was made on 2 July 1859 by John C Wise, John La Mountain and O A Gager, who covered 1120 miles (1800 km) from St Louis to Henderson, NY.

The first telegraph message transmitted from the air (and then relayed to the President of the United States) was keyed out by an official telegraph operator who accompanied the flamboyant showman, Thaddeus Sobieski Constantine Lowe, during a tethered demonstration flight in the balloon *Enterprise*, on 18 June 1861.

Inflating the balloon *Intrepid* during the American Civil War Battle of Fair Oaks, 31 May–1 June 1862 (USAF)

The first American Army Balloon Corps was formed on 1 October 1861 with a complement of 50 men under the command of Thaddeus S C Lowe who, following his demonstration flight, had been made Chief Aeronaut of the Army of the Potomac. The Corps had originally five balloons, the *Constitution, Intrepid, Union, United States* and *Washington*. Two more, the *Excelsior* and *Eagle*, entered service early in 1862. They were used for reconnaissance and artillery direction. The Corps was disbanded in mid-1863, almost two years before the end of the American Civil War.

The world's first aircraft carrier (defined as a waterborne craft used to tether, transport or launch an aircraft) was the *G W Parke Custis*, a coal-barge converted during the American Civil War in 1861 under the direction of Thaddeus S C Lowe for the transport and towing of observation balloons. The *G W Parke Custis* entered service with General McClellan's Army of the Potomac in November 1861, frequently towing balloons on the Potomac River for the observation of the opposing Confederate forces.

The first balloon ascent in Australia was made on 29 March 1858 by two men named Brown and Dean in a hydrogen balloon, the *Australasian*, from Cremorne Gardens, Melbourne. (In 1851 a Dr William Bland, 27 years after he had been transported from India to Australia for killing a ship's purser in a duel, attempted to produce a powered balloon, but there is no evidence that this ever flew.)

The first military use of balloons in an international war outside Europe was by the Brazilian Marquis de Caxias during the Paraguayan War of 1864–70. This atrocious conflict, which committed the combined forces of Brazil, Argentina and Uruguay against landlocked Paraguay, brought total disaster to the latter nation whose dictator, Francisco Solano López, ordered mass killings among his own people in a savage attempt to compel them towards victory. In the event Brazil occupied Paraguay until 1876; of about 250000 Paraguayan male nationals before the war, only 28000 survived in 1871.

The first major balloon operation was carried out during the Franco-Prussian War of 1870–1. The Prussian Army had surrounded Paris and had cut off the city from the rest of France. Inside the city were a few skilled balloonists and material for balloon-making. In an attempt to get despatches out of Paris, Jules Duruog ascended in a balloon on 23 September 1870. He flew over the Prussian camp and landed at Evreux three hours later, followed by Gaston Tissandier, Eugène Godard and Gabriel Mangin, who were all fired on. Meanwhile inside Paris other balloons were being made from available material and sailors from the French Navy were being trained as pilots. Balloon ascents carried on until 28 January 1871, by which time 66 flights had been made, carrying about 110 passengers in addition to the pilots, $2\frac{1}{2}$–3 million letters, and carrier pigeons to fly back to Paris with despatches. In mid-October 1870 a chemist, M Barreswil, suggested the use of microphotography to allow each pigeon to carry a large number of messages. On 18 November **the first official pigeon post** was introduced between Tours and Paris. This microphotography system was reintroduced during the Second World War as the Airgraph service, coping with large volumes of forces and civilian airmail.

The first practical development of balloons in the British Army dates from 1878 when the **first 'air estimates'** by the War Office allocated the sum of £150 for the construction of a balloon. Captain J L B Templer of the Middlesex Militia (later KRRC(M)) and Captain H P Lee, RE, were appointed to carry out the necessary development work. Although Captain Templer was thus the **first British Air Commander** and an aeronaut in his own right (and the owner of the balloon *Crusader*, which became the **first balloon used by the British Army** in 1879), the **first two aeronauts in the British Army** were Lieutenant (later Captain) G E Grover, RE, and Captain F Beaumont, RE, who were attached as aeronauts to the Federal Army during the American Civil War from 1862. The **first British Army balloon,** a coal-gas balloon named *Pioneer*, was made during 1879, costing £71 from the £150 appropriation, and had a capacity of 10000 ft³ (283·2 m³).

The first balloon ascent in Canada was made on 31 July 1879 by a hydrogen balloon manned by Richard Cowan, Charles Grimley and Charles Page at Montreal.

The first military use of a man-carrying balloon in Britain was that by a balloon detachment during military manoeuvres at Aldershot, Hants, on 24 June 1880. A balloon detachment accompanied the British military expedition to Bechuanaland, arriving at Cape

US Army hydrogen balloon at Fort Myer, Virginia, summer 1908

Town on 19 December 1884, and another accompanied the expeditionary force to the Sudan, departing from Britain on 15 February 1885.

The first attempt to carry out an exploration of the Arctic by free balloon was made on 11 July 1897, when Salomon August Andrée and two companions took off from Danes Island, Spitzbergen. Their 160 000 ft³ (4531 m³) capacity balloon had a sail attached to a complicated arrangement of drag ropes, with which it was hoped to steer the craft. Nothing was known of the explorers' fate until their bodies were discovered on White Island, Franz Josef Land, on 6 August 1930.

The first official balloon race in Great Britain, organised by the Aero Club, took place on 7 July 1906. Seven balloons competed, taking off from the grounds of the Ranelagh Club at Barn Elms, London. Winner of the event was Frank Hedges Butler, accompanied by Colonel J C and Mrs Capper.

The first international balloon race, and also the first of the balloon races for the Gordon Bennett Trophy, attracted an entry of 16 balloons. Flown from the gardens of the Tuileries, Paris, on 30 September 1906, it was won by Lt Frank P Lahm of the US Army who covered a distance of 402 miles (647 km) before landing at Fylingdales Moor, near Whitby, Yorkshire.

The first air crossing of the North Sea was made during 12–13 October 1907 by the hydrogen balloon *Mammoth* manned by Monsieur A F Gaudron (French aeronaut) accompanied by two others. They ascended from Crystal Palace, London, and landed at Brackan on the shore of Lake Vänern in Sweden. The straight-line distance flown was about 720 miles (1160 km).

The first international balloon race to be held in Great Britain, on 30 May 1908, was flown from the grounds of the Hurlingham Club, Fulham, London. Thirty balloons competed, representing five European nations.

The period 1895–1914 has been termed 'the Golden Age' of ballooning. The science and craft of ballooning, for such it had become with the formation of military balloon units in many parts of the world, was now to be joined by ballooning as a respectable sport and recreation. The showmen-aeronauts began to disappear; stunt flights gave way to organised competition. This was becoming the age of the motor car and the aeroplane. International sport blossomed under the watchful eye of respected clubs and societies. Undoubtedly the greatest and longest-lived international ballooning contest was the James Gordon Bennett Trophy, a competition which continued to be held almost every year from 1906 until the Second World War. (See page 26–27). While altitude record-breaking provided a spur for human achievement, scientific research of the atmosphere provided the necessary finance.

The great days of ballooning, in Edwardian England: a balloon ascent during a fete at the Crystal Palace, London

The first ratified altitude record for balloons was that achieved on 30 June 1901 by Professors Berson and Suring of the Berliner Verein für Luftschiffahrt who attained a height of 35 435 ft (10 800 m). At the time of this record's ratification there was much controversy with those who still firmly believed that James Glaisher had achieved a height of 37 000 ft (11 275 m) on 5 September 1862; as instrumentation to confirm this altitude with any chance of accuracy did not exist at the time, ratification of the Berson and Suring record was upheld; this record remained unbroken for 30 years (although exceeded on a number of occasions by aeroplanes). Berson and Suring's record was beaten in 1931 when the Swiss physicist Auguste Piccard, carried in a sealed capsule suspended beneath a balloon, made **the first balloon flight into the stratosphere** with an altitude of 50 135 ft (15 281 m). In the following year he increased this to 53 153 ft (16 201 m). On 11 November 1935 Captain Orvil Anderson and Captain Albert Stevens of the USA attained an altitude of 72 395 ft (22 066 m) in a balloon in which they ascended from a point 11 miles (17 km) southwest of Rapid City, South Dakota, and landed 12 miles (19 km) south of White Lake, South Dakota.

The first operational use of intercontinental bomb-carrying balloons was made on 3 November 1944, when the Japanese initiated an assault on the United States. An ingenious constant-altitude device was intended to ensure that the balloon remained aloft in the prevailing jet stream which carried the balloons 6200 miles (9978 km) across the Pacific Ocean. Each carried a payload of one 33 lb (15 kg) anti-personnel bomb and two incendiary weapons. More than 9000 of these balloons were launched, and it is estimated that approximately 1000 completed the crossing. Because of a self-destruct device, there were only 285 recorded incidents as a result of their use, and only six persons are known to have been killed by them.

The first ratified altitude record for a manned balloon of over 100 000 ft (30 480 m) was achieved by Major David G Simons, a medical officer of the US Air Force, who reached an altitude of 101 516 ft (30 942 m) on 19–20 August 1957 in a 3 000 000 ft³ (84 950 m³) balloon *AF-WRI-1*. He took off from Crosby, Minnesota, on 19 August to gather scientific data in the stratosphere and landed at Frederick, South Dakota, the following day.

The current world altitude record for manned free balloons is held by Commander Malcolm D Ross of the United States Navy Reserve who, on 4 May 1961, ascended over the Gulf of Mexico to an altitude of 113 739·9 ft (34 668 m) in the Lee Lewis Memorial Winzen Research balloon.

The current world altitude record for manned hot-air balloons was set by Julian Nott and Felix Pole of Great Britain on 25 January 1974. Flown from Bhopal, central India, their 375 000 ft³ (10 618 m³) capacity balloon *Daffodil II* (G-BBGN), attained an altitude of 44 550 ft (13 580 m).

The world's largest hot-air balloon, the 500 000 ft³ (14 158 m³) *Gerard A Heineken*, built by Cameron Balloons of Bristol, Avon, flew for the first time on 19 August 1974. 12 passengers were carried on this maiden flight in a unique two-tier basket.

Gerard A Heineken, the world's largest hot-air balloon, with accommodation for 30 persons. Holder of the world duration record for hot-air balloons, at 18 h 56 min

DIRIGIBLES (NAVIGABLE AIRSHIPS)

The arrival of the free balloon in 1783 represented a culmination of man's attempts to become airborne, albeit without a reliable means of navigation. The scarcely predictable nature of this means of travel was a frustration, and not many years passed before efforts were made to steer, and ultimately to propel, balloons at speeds greater and in directions other than that of the wind. Although the first elongated and theoretically steerable balloon was attributed to the Frenchman, Lieutenant Jean-Baptiste Marie Meusnier (1754–93), who published a design of such a craft in 1784, it was not until 1852 that a powered dirigible first carried a man into the air.

The world's first powered, manned dirigible made its first flight on 24 September 1852, when the Frenchman, Henri Giffard, rose in a steam-powered balloon from the Paris Hippodrome and travelled approximately 17 miles (27 km) to Trappes, at an average speed of 5 mph (8 km/h). The envelope was 144 ft (43·89 m) in length and had a capacity of 88 000 ft³ (2492 m³); the steam engine developed about 3 hp and drove an 11 ft (3·35 m) diameter three-blade propeller.

The world's first dirigible to be powered by an internal combustion engine was that built and flown by Paul Haenlein in Vienna, Austria, in 1872. Approximately 164 ft (50 m) in length and 29 ft 6 in (9 m) maximum diameter, the 85 000 ft³ (2407 m³) craft was powered by a Lenoir-type four-cylinder gas engine which consumed gas from the envelope. The engine developed about 5 hp, turning a 15 ft (4·57 m) diameter propeller at about 40 rpm, using 250 ft³ (7·08 m³) of gas per hour. Only tethered flights were made and lack of capital prevented further development.

The world's first fully controllable powered dirigible was La France, an electric-powered craft which, flown by Captain Charles Renard and Lieutenant Arthur Krebs of the French Corps of Engineers, took off on 9 August 1884 from Chalais-Meudon, France, flew a circular course of about 5 miles (8 km), returned to their point of departure and landed safely. The 9 hp Gramme electric motor drove a 23 ft (7·01 m) four-blade wooden tractor propeller. A maximum speed of 14½ mph (23·5 km/h) was achieved during the 23 min flight.

The first successful use of a petrol engine in a dirigible was by the German Dr Karl Wölfert who designed and built a small balloon to which he fitted a 2 hp single-cylinder Daimler engine in 1888. Its **first flight** was carried out at Seelberg, Germany, on Sunday, 12 August that year, probably flown by a young mechanic named Michaël.

The first flight of a Zeppelin dirigible was made by Count Ferdinand von Zeppelin's LZ1 on 2 July 1900 carrying five people from its floating hangar on Lake Constance near Friedrichshafen. The flight lasted about 20 min.

The first dirigible to fly round the Eiffel Tower and so win a 100 000 franc prize was the No 6, built and piloted by Santos-Dumont, on 19 October 1901. With an overall length of 108 ft (33 m), maximum diameter of 19 ft 6 in (6 m) and capacity of 22 200 ft³ (630 m³), it was powered by a 20 hp Buchet/Santos-Dumont water-cooled petrol engine, driving a two-blade propeller of 13 ft (4 m) diameter.

The first practical dirigible was designed by the Lebaudy brothers and first flew on 12 November 1903. The flight of 38 miles (61 km) was from Moisson to the Champ-de-Mars, Paris. This was the **first controlled air journey in history.**

Zeppelin LZ 1 over Lake Constance, 2 July 1900
(Deutsches Museum)

The first British Army airship, Dirigible No
1 (popularly known as *Nulli Secundus*), was first
flown on 10 September 1907 with three occupants: Colonel John Capper, RE, pilot; Captain
W A de C King, Adjutant of the British Army
Balloon School; Mr Samuel Cody, 'in charge
of the engine'. The engine was a 50 hp Antoinette. The airship was 122 ft (37 m) long, 26 ft
(8 m) in diameter and had a capacity of 55 000 ft³
(1555 m³). The second and third Army airships
were *Beta* (35 hp Green engine) and *Gamma*
(80 hp Green engine) respectively.

**The first occasion on which four people lost
their lives in an air accident** was on 25 September 1909 when the French dirigible *Repub-
lique* lost a propeller which pierced the gasbag;
the craft fell from 400 ft (122 m) at Avrilly, near
Moulins, the crew of four being killed.

**The first occasion on which five people lost
their lives in an air accident** was on 13 July
1910 when a German non-rigid dirigible, of the
Erbslöh type, suffered an explosion of the gasbag and fell from 920 ft (280 m) near Opladen,
Germany. The crew of five, including Oscar
Erbslöh, were killed.

**The first flight of an airship from England
to France** was achieved by E T Willows in the
Willows III on 4 November 1910.

The first passenger services by air were
operated between cities in Germany by five
Zeppelin airships in the years 1910–14. More
than 35,000 people were carried without injury.

The first dirigible of the US Navy to fly was
the DN-1 (A1) which was acquired under contract on 1 June 1915. As originally built with two
engines it was too heavy to leave the ground;
after redesign with only one engine it made its
first of three flights in April 1917 at Pensacola,
Florida. It was subsequently damaged and not
repaired.

The first successful dirigible of the US Navy
was the Goodyear FB-1 acquired under contract
on 14 March 1917 and first flown from Chicago,
Illinois, to Wingfoot Lake, near Akron, Ohio,
on 30 May 1917.

**The first airship crossing, and first two-way
crossing of the Atlantic** were achieved by the
British airship R-34 between 2 and 6 July (west-
ward) and 10 and 13 July (eastward), 1919. Commanded by Squadron Leader G H Scott, with a
crew of 30, the R-34 set out from East Fortune,

Scotland, and flew to New York, returning
afterwards to Pulham, Norfolk, England. The
total distance covered, 6330 miles (10 187 km)
in 183 h 8 min, constituted a world record for
airships. The later British airship R-38 was the
largest in the world, with a length of 695 ft
(212 m), maximum diameter of 85 ft (26 m) and
capacity of 2 740 000 ft³ (77 600 m³). It was
powered by six engines giving a total output
of 2100 hp. (The R-38 was destroyed when it
broke up over Hull, England, on 24 August
1921. It was to have been sold to the United
States and at the time of the disaster there were
15 Americans aboard in addition to the crew of
34. All the Americans and 29 of the British lost
their lives.)

**The first helium-filled American rigid air-
ship** was the Zeppelin-type ZR-1 *Shenandoah*,
which first flew on 4 September 1923 at Lakehurst, New Jersey, USA. On 3 September 1925
it was destroyed in a storm over Caldwell, Ohio,
with heavy loss of life.

The first airship flight over the North Pole
was made by the Italian-built N-class semi-rigid
airship N.1, subsequently named *Norge* by
Roald Amundsen who bought the airship for
Arctic exploration. During the period 11–14
May 1926, the *Norge* was flown from Spitzbergen to Teller, Alaska. Among the distinguished crew were Amundsen, Umberto Nobile
and Lincoln Ellsworth, who dropped Norwegian, Italian and American flags at the Pole on
12 May.

**The first round-the-world flight by an air-
ship** was made by the German Zeppelin LZ-127
Graf Zeppelin, in the period 8–29 August 1929.

The R-100

Most successful of all the passenger-carrying airships, the *Graf* had flown well over a million miles and had carried a total of some 13 100 passengers before being scrapped at the beginning of the Second World War.

The last commercial airships to be developed by Great Britain were the R-100 and R-101. The latter crashed on 5 October 1930 at Beauvais, France, on a flight from Cardington, Bedfordshire, England, to Egypt and India. The accident, which destroyed the airship and killed 48 of the 54 occupants (including Lord Thompson, Secretary of State for Air, and Major-General Sir Sefton Brancker, Director of Civil Aviation), brought to an end the development of passenger-carrying airships in Great Britain. The R-100 was designed by Barnes Neville Wallis (later Sir Barnes) for the Airship Guaran-

tee Company. It was for this craft that he originated his unique geodetic form of basic airframe structure, used later in the construction of Vickers Wellesley and Wellington bomber aircraft. In a test on 16 January 1930 the R-100 achieved a speed of $81\frac{1}{2}$ mph (131 km/h), making it **the fastest airship in the world.** Later in 1930 this craft completed successfully a double Atlantic crossing but, following the disaster to the R-101 in October, was scrapped.

The last major airship disaster involved the destruction of the German *Hindenburg*, then the world's largest airship, on 6 May 1937. It was destroyed by fire when approaching its moorings at Lakehurst, New Jersey, USA, after a flight from Frankfurt, Germany. Thirty-five of the 97 occupants were killed in the fire which engulfed the huge craft and which was attributed

LZ-127 *Graf Zeppelin*, first airship to fly around the world

Norge, first airship to fly over the North Pole

to the use of hydrogen – the only gas available to Germany owing to the United States' refusal to supply commercial quantities of helium.

The first airship built in Britain following the R-101 disaster made its first flight at Cardington, Bedfordshire on 19 July 1951. This was a small airship named *Bournemouth*, built by the Airship Club of Great Britain under the leadership of Lord Ventry.

Largest current airship fleet is operated by the Goodyear Tire & Rubber Company of Akron, Ohio, USA. It comprises four non-rigid airships, the *Mayflower III* (gross volume 147 300 ft³ (4171 m³)), *America*, *Columbia III* and *Europa* (gross volume of each 202 700 ft³ (5739 m³)). This last airship, first flown on 8 March 1972 at

Cardington, England, and now based in Rome, Italy, is made of two-ply Neoprene-coated Dacron and, like other Goodyear airships, is helium-filled. It carries a pilot and six passengers in an underslung gondola and is powered by two 210 hp Continental piston-engines which give it a cruising speed of 35–40 mph (56–64 km/h). Endurance is from 10 to 23 h. The *Europa* is 192 ft 6 in (58·67 m) long, and 50 ft (15·24 m) wide. On each side of the envelope is a four-colour sign, 105 ft (32·00 m) long and 24 ft 6 in (7·47 m) high, made up of 3780 lamps to flash static or animated messages. Newest airship of the fleet is the *Columbia III*, which entered service in 1975.

The world's first hot-air airship (G-BAMK), built by Cameron Balloons Ltd of Bristol, made

Demonstration of advertising lights on the side of the Goodyear airship *Europa*

its maiden flight near Wantage, Berkshire, on 4 January 1973. It is 100 ft (30·5 m) long, with a maximum diameter of 45 ft (13·72 m), and is propelled by a converted Volkswagen engine developing 45 hp, which gives a maximum speed of 17 mph (27·5 km/h). The lightweight nylon fabric envelope has a gross capacity of 96 000 ft^3 (2718 m^3) and is inflated by hot air generated by a propane gas burner carried in the lightweight tubular-metal gondola.

THE JAMES GORDON BENNETT INTERNATIONAL BALLOON RACE TROPHY

Without doubt the most famous international balloon contest was the Gordon Bennett contest for a trophy and an annual prize of 12 500 francs presented by the expatriate American newspaper magnate, James Gordon Bennett, in 1905 and first contested in 1906. Principal results were as follows:

Date	Starting-point and number of Starters	Winners (Nationality, balloon, qualifying destination and distance)		Remarks
1906	Tuileries, Paris, France (16)	Lieutenant Frank P Lahm, US Army; *United States*; Fylingdales, Yorkshire, England	402 miles (647 km)	Only three balloons with their crews managed to cross the English Channel.
1907	St Louis, Missouri, USA	Oscar Erbslöh, Germany; *Pommern*	849 miles (1367 km)	
1908	Berlin, Germany (23)	Colonel Schaeck, Switzerland; *Helvetia*; 40 km N of Molde, Norway	753 miles (1212 km)	Many balloons came down in the North Sea.
1909	Zürich, Switzerland (17)	E W Mix, USA, *America II*; Ostrolenka, Poland	590 miles (950 km)	
1910	St Louis, Missouri, USA	Allan R Hawley, USA; *America II*; Lake Tschotogama, Quebec, Canada	1171·13 miles (1884·75 km)	Established a new American distance record.
1911	Kansas City, Kansas, USA	O Gericke, Germany; Holcombe, Wisconsin, USA	468·2 miles (753·5 km)	
1912	Stuttgart, Germany	A Bienaimé, France; *Picardi*; Ryasan, near Moscow, USSR	1361 miles (2191 km)	One of the longest balloon races in history.
1913	Tuileries, Paris, France	Ralph Upson, USA; *Goodyear*; Bempton, Yorkshire, England	384 miles (618 km)	
1920	Birmingham, Alabama, USA (8)	Lieutenant Ernest E Demuyter, Belgium; *Belgica*; Lake Champlain, Vermont, NY, USA	1094 miles (1760 km)	
1921	Brussels, Belgium (14)	Captain Paul Armbruster, Switzerland; Lambay Island, Ireland	476 miles (766 km)	
1922	Geneva, Switzerland (18)	Lieutenant E Demuyter, Belgium; *Belgica*; Oknitsa, Romania	853 miles (1372 km)	
1923	Brussels, Belgium (17)	Lieutenant E Demuyter, Belgium; *Belgica*; Skillingaryd, Sweden	994 miles (1600 km)	Race started in thunderstorm, met with disaster, five aeronauts killed, five injured.
1924	Brussels, Belgium	Lieutenant E Demuyter, Belgium; *Belgica*; St Abbs Head, Berwick, Scotland	466 miles (750 km)	This, the third consecutive win by Belgium, would have qualified that country to retain the trophy outright. The Belgians, however, sportingly announced its renewal for further competition.

Date	Starting-point and number of Starters	Winners (Nationality, balloon, qualifying destination and distance)		Remarks

SECOND TROPHY SERIES

1925	Brussels, Belgium	Lieutenant Veenstra, Belgium; *Prince Leopold*	841 miles (1354 km)	In this race Van Orman (US Balloon *Goodyear III*) performed the singular feat of landing on the bridge of a ship at sea; the American balloon, *Elsie*, was hit by a train at Étaples.
1926	Antwerp, Belgium	W T Van Orman, USA; *Goodyear III*; Sölvesborg, Sweden	535 miles (861 km)	
1927	Dearborn, Michigan, USA (15)	E J Hill, USA; *Detroit*; Baxley, Georgia	745 miles (1199 km)	
1928	Detroit, Michigan, USA (12)	Captain W E Kepner, USA; *US Army*; Kenbridge, Virginia	460 miles (740·3 km)	Once again the trophy was re-presented for competition, this time by the USA.

THIRD TROPHY SERIES

| 1929 | St Louis, Missouri, USA | W T Van Orman, USA; Troy, Ohio | 339 miles (545 km) | |
| 1930 | Cleveland, Ohio USA | W T Van Orman, USA; North Canton, Massachusetts | 539 miles (867 km) | No Gordon Bennett Trophy Race was held in 1931 and a fourth series of races was started in 1932. |

FOURTH TROPHY SERIES

1932	Basle, Switzerland	Lieutenant-Commander T G Settle, USA; *US Navy*: Vilna, Lithuania	954 miles (1536 km)	New series established under Swiss administration.
1933	Chicago, Illinois, USA	Z J Burzynski, Poland; Quebec, Canada	846 miles (1361 km)	
1934	Warsaw, Poland	Hynek, Poland; *Kosciuszko*; Finland (location not recorded)	827 miles (1331 km)	Landing area of winning balloon probably near Savonlinna, Finland.
1935	Warsaw, Poland	Z J Burzynski, Poland; *Polonia II*; near Leningrad, USSR	1025 miles (1650 km)	The leading balloons all landed in a desperately remote area and were not retrieved for a fortnight after landing. Burzynski's flight established a world record for balloons of 1601–2200 m^3.
1.9.36	Warsaw, Poland	Ernest Demuyter, Belgium; *Belgica*; Miedlesza, USSR	1066·15 miles (1715·8 km)	The veteran Demuyter's flight constituted a world distance record in four balloon categories.
1937	Brussels, Belgium	Ernest Demuyter, Belgium; *Belgica*; Tukumo, Lithuania	889 miles (1430 km)	
1938	Liège, Belgium	Janusz, Poland; *L.O.P.P.*; Trojan, Bulgaria	1013 miles (1630 km)	

Lilienthal flying one of his monoplane hang gliders,
1894 (Deutsches Museum)

SECTION 3

PIONEERS OF FLIGHT

Less than fifty years after Roger Bacon described the flying machine devised by his very prudent friend, the great Venetian merchant traveller Marco Polo told a strange tale about something he had seen during his journeyings in China. A translation of his report appears in *The Description of the World*, edited by A C Moule and P Pelliot and published in London in 1938; it goes like this:

And so we will tell you how, when any ship must go on a voyage, they prove whether its business will go well or ill on that voyage. The men of the ship indeed will have a hurdle, that is a grating, of withies, and at each corner and side of the hurdle will be tied a cord, so that there will be eight cords, and they will all be tied at the other end with a long rope. Again they will find someone stupid or drunken and will bind him on the hurdle; for no wise man nor undepraved would expose himself to that danger. And this is done when a strong wind prevails. They indeed set up the hurdle opposite the wind, and the wind lifts the hurdle and carries it into the sky and the men hold on by the long rope. And if while it is in the air the hurdle leans towards the way of the wind, they pull the rope to them a little and then the hurdle is set upright, and they let out some rope and the hurdle rises. . . . The proof is made in this way, namely that if the hurdle going straight up makes for the sky, they say that the ship for which that proof has been made will make a quick and prosperous voyage, and all the merchants run together to her for the sake of sailing and going with her. And if the hurdle has not been able to go up, no merchant will be willing to enter the ship for which the proof was made, because they say that she could not finish her voyage and many ills would oppress her. And so that ship stays in port that year.

Japanese man-lifting kite, *ca* 1860

Pennon-kite from Conrad Kyeser's *Bellifortis*, 1405

We know now that what Marco Polo had seen was a kite, then unknown in Europe, Anyone brave, stupid or unfortunate enough to take a trip on a manlifter enjoyed the kind of view of which other men could only dream; but kites were temperamental fliers, and there were many occasions on which the operators of the ropes shrugged their shoulders and said the equivalent of 'Poor old Chang. It's no good sailing today. Let's go for a drink to his memory.'

Such kites, being intended for purely practical purposes, were huge and plain. The earliest known account of their use concerns the Chinese General Han Hsin who, about 200 BC, flew an unmanned kite over an enemy palace so that he could measure accurately the distance between his positions and the walls of the building, by the length of rope let out. He then knew how long to make the tunnel by which his troops were to enter the palace.

In Europe, too, kites had a military origin. Trajan's Column, in Rome, depicts the form of standard carried into action by Dacian armies, and adopted by the Romans after the final defeat of Dacia in AD 105. On the lines of a modern windsock, used to indicate the wind direction at an airfield, it consisted of a hollow representation of a dragon's head, mounted on a pole and with a fabric tube streaming out behind it to form the dragon's body. Highly coloured, writhing in the wind, hissing as it did so and, perhaps, breathing out smoke or fire, the standard, or Draco, helped to inspire fear in enemies. By the 14th century the pole support had been superseded by a long rope, and some of the windsocks had sprouted wings to provide lift and a degree of stability. A drawing which dates from 1326–7 shows one of them being used by three soldiers to drop a bomb on an enemy city.

The modern kite owes more to the Chinese manlifter than to windsocks. And the century-long story of the eventual conquest of the air begins with such a kite, in the year 1804.

The setting was the Yorkshire estate of Sir George Cayley, Baronet (1773–1857), who is regarded universally as the 'Father of Aerial Navigation'. Back in 1799, this man had taken a small silver disc and engraved it with the first known design for a fixed-wing aeroplane. Still only twenty-six years old, he devised the key to eventual success by suggesting that the lifting system should be left to do its own job efficiently, instead of having also to provide propulsion by flapping. If we make allowances for his obsession with paddles rather than a propeller as the medium for producing thrust, his design embodied all the features of a modern aeroplane, with a fixed and cambered wing, set at a modest angle of attack, a man-carrying body, and movable cruciform tail surfaces.

When Cayley built these deceptively simple ideas into what – despite the ancient Egyptian model – is regarded as the world's first successful model glider, in 1804, he used as the wing what any young person would recognise as a kite-form. It was followed by a successful full-scale glider five years later and, ultimately, by the paddle-equipped glider in which the eighty-year-old Cayley is said to have climaxed his career by flying his aged, and highly reluctant, coachman over a small valley in 1853.

A handful of almost predictable steps led from there to the triumph of Wilbur and Orville Wright, at Kitty Hawk, in December 1903. They spanned half a century mainly because nobody had yet invented a light-weight, powerful engine that would lift into the air a flying-machine based on Cayley's principles.

Most inspired of all the products of the pioneers who followed Cayley was the first of them. We can still see it exactly as its designer, William Samuel Henson (1812–88), conceived it; because his 20 ft span model of the Aerial Steam Carriage is a treasured exhibit in London's Science Museum. It is a monoplane, with enclosed cabin and tricycle undercarriage. Given a power plant more efficient than its steam engine, and perhaps a few degrees of wing dihedral, it looks as if it might have flown, and the course of aviation history would have been changed. Instead, when the model failed to fly, Henson was ridiculed by the sceptics, who poured scorn on drawings of full-scale Steam Carriages in flight over places like the Pyramids and the Taj Mahal.

It was left to a French naval officer, Félix Du Temple de la Croix, to fly a clockwork model in 1857 and then to scale it up into a sweptforward-wing monoplane which lifted a young sailor into the air

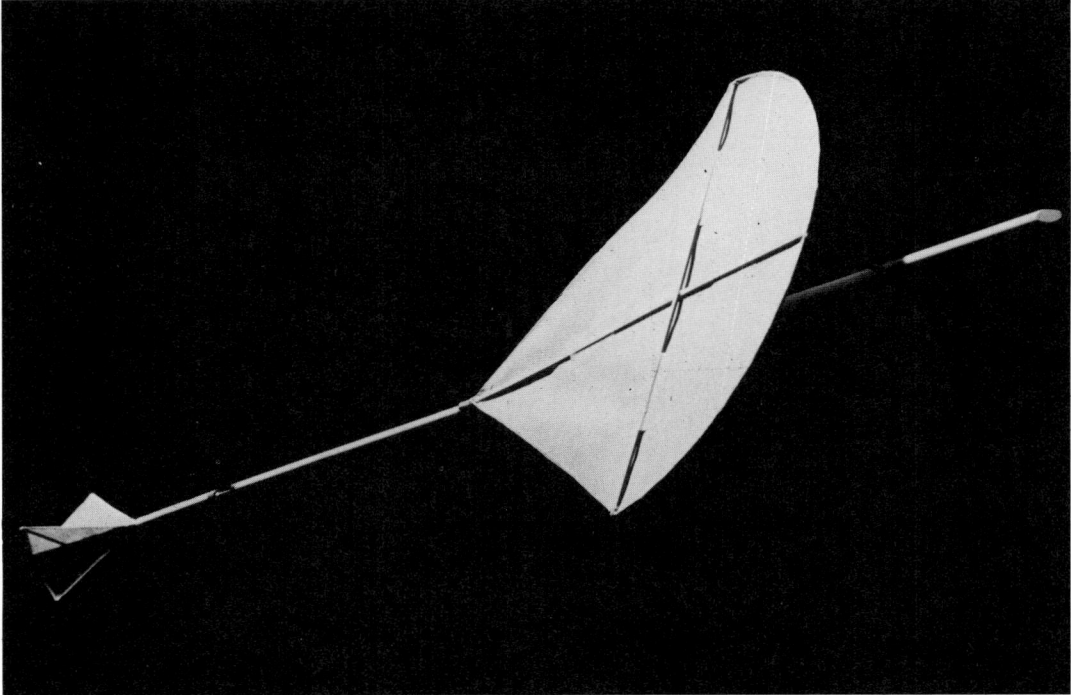

Reproduction of Cayley's 1804 model glider in the Qantas model collection

at Brest 17 years later. It was not a proper flight, as the aircraft made only a short hop after gathering speed down a ramp. The huge steam-powered monoplane of Alexander Mozhaisky made a similar ramp-launched leap through the air in Russia in 1884. Ten years later, in Britain, Sir Hiram Maxim's 3·5 ton biplane lifted itself briefly and unintentionally into the air, from level ground, also under steam-power, but was hardly a practical aeroplane.

Realising that it was now possible to design and build a thoroughly efficient airframe, but that no suitable power plant was available, Otto Lilienthal reverted to powerless gliding flight in Germany. By making more than 2000 successful glides in 1890–96, he inspired other glider builder/pilots like the American Octave Chanute (1832–1910) and the British Percy S Pilcher (1866–99). By a tragic coincidence, both Lilienthal and Pilcher died in glider crashes as they were about to fit small, practical engines to their best designs.

So, the race to be first to achieve a powered, sustained, controlled flight was eventually between three Americans, Dr Samuel Pierpont Langley and the brothers Wilbur and Orville Wright. The odds favoured Langley when he launched his steam-powered pilotless model *Aerodrome No. 5* over the Potomac River on 6 May 1896, for a sustained flight lasting more than 1 min; but two attempts to fly a full-scale piloted version were to prove unsuccessful in the autumn of 1903.

Over in Germany, Carl Jatho (1873–1933) had greater success with a lesser aeroplane. His first machine was a biplane only 11 ft 10 in (3·60 m) long, with the elevator mounted triplane-like on top. There were two vertical rudders between the wings, a rubber-tyred undercarriage, and a large propeller driven by a 9·5 hp Buchet engine. Wing area was 581 ft² (54 m²) and the machine weighed 562 lb (255 kg).

Best flight achieved with this aircraft was 59 ft (18 m), at a height of about 2 ft 6 in (75 cm), on 18 August 1903. Jatho redesigned it as a monoplane, still with a top elevator, to make it easier to control in windy conditions. His second prototype had a reduced wing area of 387·5 ft² (36 m²) and weight of only 408 lb (185 kg). When tested in the autumn of 1903, it made a number of flights of up to 196 ft (60 m), at a height of about 11 ft (3·5 m). It was to be several years before he flew further, in a better design with more power; but it is worth pointing out that his longest hops far exceeded the first of the four flights by the Wright brothers, which they preceded. So Jatho, like the other pioneers who kept enthusiasm alive between the major 'milestones' of Cayley and the Wrights, should never be forgotten.

The first man in the world to identify and correctly record the parameters of heavier-than-air flight was the Englishman, Sir George Cayley, sixth Baronet (born 27 December 1773; died 15 December 1857). Sir George Cayley succeeded to the Baronetcy in a long and distinguished line of Cayleys whose origins are traceable back to Sir Hugo de Cayly, Knight, of Owby, who lived early in the 12th century. Sir William Cayley was created first Baronet by Charles I on 26 April 1661 for services in the Civil War. The present Baronet, Sir Kenelm Henry Ernest Cayley, tenth Baronet, of Brompton, Yorkshire, was born on 24 September 1896. It was at Brompton Hall, near Scarborough, 100 years before, that young George Cayley had carried out some of his early experiments with model aeroplanes. The following is a list of notable 'firsts' achieved by this remarkable scientist:

(a) He first set down the mathematical principles of heavier-than-air flight (i.e. lift, thrust and drag).

(b) He was the first to make use of models for flying research, among them a simple glider – the first monoplane with fixed wing amidships, and fuselage terminating in vertical and horizontal tail surfaces; this was constructed in 1804.

(c) He was the first to draw attention to the importance of streamlining (in his definition of 'drag').

(d) He was the first to suggest the benefits of biplanes and triplanes to provide increased lift with minimum weight.

(e) He was first to construct and fly a man-carrying glider (see page 32).

(f) He was the first to demonstrate the means by which a curved 'aerofoil' provided 'lift' by creating reduced pressure over the upper surface when moved through the air.

(g) He was the first to suggest the use of an internal-combustion engine for aeroplanes and constructed a model gunpowder engine in the absence of low-flash-point fuel oil.

Model of Cayley's triplane glider of 1849
(Qantas Collection)

The first person to be carried aloft in a heavier-than-air craft in sustained (gliding) flight was a ten-year-old boy who became airborne in a glider constructed by Sir George Cayley at Brompton Hall, near Scarborough, Yorkshire, in 1849. The glider became airborne after being towed by manpower down a hill against a light breeze.

The first man to be carried aloft in a heavier-than-air craft, but not in control of its flight, was Sir George Cayley's coachman at Brompton Hall, reputedly in 1853. A witness of the event stated that after he had landed the coachman struggled clear and shouted 'Please, Sir George, I wish to give notice. I was hired to drive, not to fly.' No record has ever been traced giving the name of either the ten-year-old or of the coachman. The decennial census of 1851, however, records the name of John Appleby as being the most probable member of Sir George's staff. With regard to the young boy, Sir George had no son or grandson of this age at the time of his experiments, so the first 'pilot' may have been a servant's son.

The first model aeroplane powered by a steam engine was that designed and made by W S Henson (1812–88) and John Stringfellow (1799–1883) at Chard, Somerset, England, in 1847. This 20 ft (6·1 m) span monoplane powered by a steam engine driving twin pusher propellers, was launched from an inclined ramp, but sustained flight was not achieved. Henson subsequently emigrated to America, but both he and Stringfellow remained interested in aeronautics and pursued experiments with powered models. Contrary to former opinion, none of

Stringfellow's powered models ever achieved sustained flight.

The first powered model aeroplane to make a successful sustained flight was built and tested in France by Félix Du Temple (1823–90) in 1857–58. It was powered successively by clockwork and steam.

The first scientist correctly to deduce the main properties (i.e. lift distribution) of a cambered aerofoil was F H Wenham (1824–1908) who built various gliders during the mid 19th century to test his theories. In collaboration with John Browning, Wenham built **the world's first wind tunnel** in 1871 for the Aeronautical Society of Great Britain.

The first aeronautical exhibition in Great Britain was staged at the Crystal Palace in 1868 by the Aeronautical Society of Great Britain. The exhibits included model engines driven by steam, oil, gas and guncotton.

The first powered man-carrying aeroplane to achieve a brief 'hop', after gaining speed down a ramp, was a monoplane with swept-forward wings built by Félix Du Temple and piloted by an unidentified young sailor, at Brest in about 1874. The power plant was a hot-air or steam engine, driving a tractor propeller.

The first man-carrying aeroplane to achieve a powered 'hop' after rising from supposedly level ground was the bat-winged *Éole* monoplane built and flown by Clément Ader (1841–1925), at Armainvilliers, France, on 9 October 1890. Powered by an 18–20 hp steam engine, the *Éole* covered about 165 ft (50 m), but never achieved sustained or controlled flight.

Replica of Félix Du Temple's aeroplane, 1874
(Qantas Collection)

Model of Ader's *Éole*

Ader's second aeroplane, the *Avion III*, was tested twice in 1897 but did not fly.

The three most outstanding pioneers of gliding flight prior to successful powered flight were the German Otto Lilienthal (1848–96), the American Dr Octave Chanute (1832–1910), and the Englishman Percy S Pilcher (1866–99). Their achievements may be summarised as follows:

Otto Lilienthal, German civil engineer, published a classic aeronautical textbook *Der Vogelflug als Grundlage der Fliegekunst* (The Flight of Birds as the Basis of Aviation) in 1899. Although he remained convinced that powered flight would ultimately by achieved by wing-flapping (i.e. in the ornithopter), Lilienthal constructed five fixed-wing monoplane gliders and two biplane gliders between 1891 and 1896. Tested near Berlin and at the Rhinower Hills near Stöllen, these gliders achieved sustained gliding flight; the pilot, usually Lilienthal himself, supported himself by his arms, holding the centre section of the glider. Thus, he could run forward and launch himself off the hills to achieve flight. During this period he achieved gliding distances ranging from 330 ft (100 m) to more than 820 ft (250 m). Although he had been experimenting with a small carbonic acid gas engine he was killed when one of his gliders crashed on the Rhinower Hills on 9 August 1896 before he could progress further with powered flight.

Octave Chanute, American railroad engineer, was born in Paris, France, on 18 February 1832. His book *Progress in Flying Machines*, published

Replica of Pilcher's *Hawk*

Lawrence Hargrave flying a box-kite, 1893

in 1894, was the first comprehensive history of heavier-than-air flight and is still regarded as a classic of aviation literature. As well as providing a valuable information service for pioneers on both sides of the Atlantic, he began designing and building improved Lilienthal-type hang-gliders in 1896. After experimenting with multi-planes fitted with up to eight pairs of pivoting wings and a top fixed surface, he evolved the classic and successful biplane configuration. Flight-testing of his gliders was performed mainly by Augustus M Herring (1867–1926), as Chanute was too old to fly himself. The Wright brothers gained early inspiration from *Progress in Flying Machines*, became close friends of Chanute, and learned from him the advantages offered by a Pratt-trussed biplane structure and, later, a catapult launching system for their wheel-less aircraft.

Percy S Pilcher, English marine engineer,

built his first glider, the *Bat*, in 1895 and flew that year on the banks of the River Clyde. Following advice by Lilienthal, as well as early practical experiments, Pilcher added a tailplane to the *Bat* and achieved numerous successful flights. This aircraft was followed by others (christened the *Beetle*, *Gull* and *Hawk*), the last of which was constructed in 1896 and included a fixed fin, a tailplane and a wheel undercarriage. It had a cambered wing with a span of 23 ft 4 in and an area of 180 ft^2 (7 m and 16·72 m^2 respectively). Pilcher had always set his sights upon powered flight and was engaged in developing a light 4 hp oil engine (probably for installation in his *Hawk*) when, having been towed off the ground by a team of horses, he crashed in the *Hawk* at Stanford Park, Market Harborough on 30 September 1899, and died two days later.

The box-kite structure was invented in 1893 by an Australian, Lawrence Hargrave (1850–

1915). This simple structure provided good lift and stability and formed the basis of early aeroplanes such as the Voisin.

The largest aeroplane to lift itself off the ground briefly in the 19th century was designed and built by Sir Hiram Maxim (1840–1916). Basically a biplane with 4000 ft² (372 m²) of lifting area, it was powered by two 180 hp steam engines and ran along a railway track 1800 ft (550 m) long which was fitted with wooden restraining guard-rails to prevent the machine from rising too high.

On 31 July 1894 during a test run the machine lifted about 2 ft (60 cm) before fouling the guardrails and coming to rest.

The first man to achieve sustained powered flight with an unmanned heavier-than-air craft was the American Samuel Pierpont Langley (born 22 August 1834 at Roxbury, Massachusetts; died 27 February 1906, at Aiken, South Carolina). Mathematician and solar radiation physicist, Langley commenced building powered model aeroplanes during the 1890s, launching them from the top of a houseboat on the Potomac River near Quantico. His 14 ft (4·25 m) span models (*Aerodrome* Nos 5 and 6) achieved sustained flights of up to 4200 ft (1280 m) during 1896 and incorporated a single steam engine mounted amidships, driving a pair of airscrews. Langley's use of the name *Aerodrome* was derived incorrectly from the Greek αερο-δρόμος (*aerodromos*) supposedly meaning 'air runner'; the word, however, is correctly defined as the location of a running event and cannot be held to mean the participant in a running event. Thus in the context of an *airfield* the word 'aerodrome', as originally applied to Hendon in Middlesex, England, is correct. In 1898 Langley was requested to continue his experiments with a $50000 State subsidy and set about the design and construction of a full-scale version. As an intermediate step he built a quarter-scale model which in August 1903 became the world's first aeroplane powered by a petrol engine to achieve sustained flight. His full-size *Aerodrome*, with a span of 48 ft (14·6 m) and powered by a 52 hp Manly-Balzer five-cylinder radial petrol engine, was completed in 1903, and attempts to fly this over the Potomac River with Charles M Manly at the controls were made on 7 October and 8 December 1903. On both occasions the aeroplane fouled the launcher and dropped into the river. In view of the success achieved by the Wright brothers immediately thereafter, the American Government withdrew its support from Langley and his project was abandoned.

The first aeroplane to achieve man-carrying, powered, sustained flight in the world was the *Flyer*, designed and constructed by the brothers Wilbur and Orville Wright, which first achieved such flight at 10.35 h on Thursday, 17 December 1903, at Kill Devil Hills, Kitty Hawk, North Carolina, with an undulating flight of 120 ft (36·5 m) in about 12 s. Three further flights were made on the same day, the longest of which covered a ground distance of 852 ft (260 cm) and lasted 59 s. It should be emphasised that these flights were the natural culmination of some four years' experimenting by the Wrights with a number of gliders, during 1899–1903. Details of the powered *Flyer* were as follows:

Wright Flyer No 1 (1903)
Wing span: 40 ft 4 in (12·3 m).
Over-all length: 21 ft 1 in (6·43 m).
Wing chord: 6 ft 6 in (1·97 m).
Wing area: 510 ft² (47·38 m²).
Empty weight: 605 lb (274 kg).
Loaded weight: Approximately 750 lb (340 kg).
Wing loading: 1·47 lb/ft² (7·2 kg/m²).
Power plant: 12 bhp four-cylinder water-cooled engine lying on its side and driving two 8 ft 6 in (2·59 m) diameter propellers by chains,

Langley's Aerodrome on its houseboat launcher, 7th October 1903 (Smithsonian Institution).

Pilotless kite used by Wilbur and Orville Wright to test their theories on control and stability, in 1900 (Smithsonian Institution)

one of which was crossed to achieve counter-rotation. Engine weight with fuel (0·33 imp. gal), approximately 200 lb (90 kg).

Speed: 30 mile/h (approximately 45 km/h).

Launching: The *Flyer* took off under its own power from a dolly which ran on two bicycle hubs along a 60 ft (18 m) wooden rail.

The first accredited sustained flight (i.e. other than a 'hop') achieved by a manned, powered aeroplane in Europe was made on 12 November 1906 by the Brazilian constructor-pilot Alberto Santos-Dumont (1873–1932), a resident of Paris, France, who flew his '*14-bis*' 722 ft (220 m) in 21·2 s; his aeroplane was in effect a tail-first box-kite powered by a 50 hp Antoinette engine, and this flight won for him the Aéro-Club de France's prize of 1500 francs for the first officially observed flight of more than 100 m. A previous flight, carried out on 23 October, covered nearly 200 ft (60 m) and had won Santos-Dumont the Archdeacon Prize of 3000 francs for the first sustained flight of more than 25 m.

Perhaps the most celebrated aviation photograph ever taken – Orville Wright making the first powered, sustained and controlled flight in history, 17 December 1903 (Smithsonian Institution)

The first powered aeroplane flight in Great Britain, though not officially recognised, was almost certainly made by Horatio Phillips (1845–1924) in a 22 hp multiplane in 1907. The aircraft had four of Phillips's unique narrow 'Venetian blind' wing-frames in tandem. It covered a distance of about 500 ft (152 m).

The first monoplane with tractor engine, enclosed fuselage, rear-mounted tail-unit and two-wheel main undercarriage with tailwheel was the Blériot VII powered by a 50 hp Antoinette engine. This was Louis Blériot's third full-size monoplane and was built during the autumn of 1907 and first flown by him at Issy-les-Moulineaux, France, on 10 November 1907. Before crashing this aeroplane on 18 December that year Blériot had achieved six flights, the longest of which was more than 1640 ft (500 m). This success confirmed to the designer that his basic configuration was sound – so much so that despite a 30-year deviation into biplane design, Blériot's basic configuration is still regarded as fundamentally conventional among propeller-driven aeroplanes of today.

The first free flight of a helicopter, with a man on board, was on 13 November 1907 near Lisieux, France. It was built by Paul Cornu and had a 24 hp Antoinette engine driving two rotors. Although this was the first free flight with a man on board, earlier, on 29 September 1907, at Douai, the Breguet-Richet Gyroplane No 1 had lifted from the ground; but four men were required to steady the machine and so Louis Breguet could not claim a free flight.

The first specification for a military aeroplane ever issued for commercial tender was drawn up by Brigadier-General James Allen, Chief Signal Officer of the US Army, on 23 December 1907. The specification (main points) was as follows:
- Drawings to scale showing general dimensions, shape, designed speed, total surface area of supporting planes, weight, description of the engine and materials.
- The flying machine should be quick and easy to assemble and should be able to be taken apart and packed for transportation.
- Must be designed to carry two persons having a combined weight of about 350 lb and sufficient fuel for a flight of 125 miles.
- Should be designed to have a speed of at least 40 mile/h in still air.
- The speed accomplished during the trial flight will be determined by taking an average of the time over a measured course of more than 5 miles, against and with the wind.
- Before acceptance a trial endurance flight will be required of at least 1 h.
- Three trials will be allowed for speed. The place for delivery to the Government and trial flights will be Fort Myer, Virginia.
- It should be designed to ascend in any country which may be encountered in field service. The starting device must be simple and transportable. It should also land in a field without requiring a specially prepared spot, and without damaging its structure.
- It should be provided with some device to permit of a safe descent in case of an accident to the propelling machine.
- It should be sufficiently simple in its construction and operation to permit an intelligent man to become proficient in its use within a reasonable length of time.

The first internationally ratified world aeroplane records of performance were those which stood at the end of 1907, both for distance covered over the ground and for flights which followed unassisted take-off from level ground. Thus although the Wright *Flyer* (1905 version) had achieved numerous observed flights across-country, these were not ratified by the FAI for world record purposes as they were, more often than not, commenced by assistance into the air by external means (a falling-weight catapult). The records established up to 31 December 1907 were thus:

12 November 1906
    Santos-Dumont ('*14-bis*')   722 ft (220 m)
26 October 1907
    Henry Farman (Voisin)   2530 ft (771 m)

The first circuit flight made in Europe was flown by Henry Farman on 13 January 1908 in his modified Voisin biplane at Issy-les-Moulineaux when he took off, circumnavigated a

Breguet-Richet Gyroplane No 1, September 1907

pylon 1625 ft (500 m) away and returned to his point of departure. By so doing Farman won the Grand Prix d'Aviation, a prize of 50000 francs offered by Henry Deutsch de la Meurthe and Ernest Archdeacon to the first pilot to cover a kilometre. The flight took 1 min 28 s and, owing to the distance taken in turning, probably covered 4875 ft (1500 m).

The first aeroplane flight in Italy was made by the French sculptor-turned aviator Léon Delagrange in a Voisin in May 1908. At this time aircraft built by the French brothers Gabriel and Charles Voisin were flown by two pilots, Henry Farman and Léon Delagrange. Henry Farman was born in England in 1874 and retained his English citizenship until 1937 when he became a naturalised Frenchman. Having turned from painting to cycling before the turn of the century, he progressed to racing Panhard motor cars and at one time owned the largest garage in Paris. Gabriel Voisin later remarked that Farman possessed considerable mechanical and manipulative skill, whereas Delagrange 'was not the sporting type' and knew nothing about running an engine. Delagrange was killed flying a Blériot monoplane on 4 January 1910. Farman, having abandoned flying to pursue the business of aeroplane manufacture died on 17 July 1958.

The first aeroplane flight in Belgium was made by Henry Farman, at Ghent, in May 1908.

The first passenger ever to fly in an aeroplane was Charles W Furnas who was taken aloft by Wilbur Wright on 14 May 1908 for a flight covering 1968 ft (600 m) of 28·6 s duration. Later the same morning Orville Wright flew Furnas for a distance of about 2½ miles (4000 m) which was covered in 3 min 40 s.

The first passenger to be carried in an aeroplane in Europe was Ernest Archdeacon, the Frenchman whose substantial prizes contributed such stimulus to European aviation, who was flown by Henry Farman on 29 May 1908.

The first American to fly after the Wright brothers was Glenn H Curtiss, who flew his *June Bug* for the first time on 20 June 1908. During this flight Curtiss covered a distance of 1266 ft (386 m) and exactly a fortnight later, on 4 July he made a flight of 5090 ft (1550 m) in 102·2 s to win the *Scientific American* trophy for the first official public flight in the United States of more than one kilometre.

Glenn H Curtiss

The first aeroplane flight in Germany was made by the Dane, J C H Ellehammer, in his triplane at Kiel on 28 June 1908. A development of this triplane was flown by **the first German pilot,** Hans Grade, at Magdeburg in October 1908.

The world's first woman passenger to fly in an aeroplane was Madame Thérèse Peltier who, on 8 July 1908, accompanied Léon Delagrange at Turin, Italy, in his Voisin for a flight of 500 ft (150 m). She soon afterwards became the first woman to fly solo, but never became a qualified pilot.

The first fatality to the occupant of a powered aeroplane occurred on 17 September 1908 at Fort Myer, Virginia, when a Wright biplane flown by Orville Wright crashed killing the passenger, Lieutenant Thomas Etholen Selfridge, US Army Signal Corps. Wright was seriously injured. The accident occurred during US Army acceptance trials of the Wright biplane and was caused by a failure in one of the propeller blades. This imbalanced the good blade, causing it to tear loose one of the wires bracing the rudder-outriggers to the wings, and so sending the aircraft crashing to the ground from about 75 ft (25 m).

The first resident Englishmen to fly in an aeroplane (albeit as passengers) were Griffith Brewer, the Hon C S Rolls, Frank Hedges Butler and Major B F S Baden-Powell, who were taken aloft in turn by Wilbur Wright in

Crash of the Wright biplane at Fort Myer, 17 September 1908

his biplane at Camp d'Auvours on 8 October 1908. Butler had founded the Aero Club of Great Britain in 1901, while Baden-Powell was Secretary of the Aeronautical Society. The 'resident' qualification is necessary here as of course the English-born, French-resident Henry Farman had been flying for more than a year by the time the four Englishmen were taken aloft by Wright.

The first officially recognised aeroplane flight in Great Britain was made by the American (later naturalised British citizen) Samuel Franklin Cody (1861–1913) in his *British Army Aeroplane No 1*, powered by a 50 hp Antoinette engine. The flight of 1390 ft (424 m) was made at Farnborough, Hants, on 16 October 1908 and ended with a crash-landing, but without physical injury to Cody.

The longest flight achieved by the end of 1908 was by Wilbur Wright on 31 December

1908, at Camp d'Auvours, where he achieved a stupendous flight of 77 miles (124 km) in 2 h 20 min. This won for him the Michelin prize of 20 000 francs – apart from breaking all his own records. A summary of the flights made by Wilbur Wright up to the end of 1908 appears on page 40, followed by a list of the more significant flights made by other aviators up to that time.

The first aerodrome to be prepared as such in England was the flying-ground between Leysdown and Shellness, Isle of Sheppey (known as 'Shellbeach'), where limited established facilities were provided. It was opened in February 1909 by the joint effort of the Aero Club of Great Britain and Short Bros Ltd.

The first sustained, powered flight by an aeroplane in the British Empire was made on 23 February 1909 by J A D McCurdy, a

Canadian, over Baddeck Bay, Nova Scotia, in his biplane *Silver Dart*, which he had designed. He had made his own first flight at Hammondsport, NY, USA, the previous December.

The first aeroplane flight in Austria was made by the Frenchman, G Legagneux, at Vienna in April 1909 in his Voisin. **The first Austrian** to fly was Igo Etrich, who flew his *Taube* at Wiener-Neustadt in November of that year. His aircraft gave its name to the type of aircraft in fairly widespread use by Germany at the beginning of the First World War.

The first cinematographer to be taken up in an aeroplane was at Centocelle, near Rome, on 24 April 1909, in a Wright biplane flown by Wilbur Wright.

The first resident Englishman to make an officially recognised aeroplane flight in England was J T C Moore-Brabazon (later Lord Brabazon of Tara) who made three sustained flights of 450, 600 and 1500 ft (130, 180 and 450 m) between 30 April and 2 May 1909 at Leysdown, Isle of Sheppey, in his Voisin biplane. He had learned to fly in France during the previous year and on 30 October 1909 won

Igo Etrich's *Taube*

Wilbur Wright

17 December 1903	Kill Devil Hills	852 ft	260 m	
9 November 1904	Dayton, Ohio	2·75 miles	(4·43 km)	Flew 105 times during 1904.
5 October 1905	Dayton, Ohio	24·2 miles	(38·9 km)	Flew 49 times during 1905.
14 May 1908	Kill Devil Hills	5 miles	(8 km)	In 7 min 29 s.
8 August 1908	Hunaudières, France	—	—	Demonstration flight 1 min 45 s.
16 September 1908	Auvours, France	—	—	Flight taking 39 min 18 s.
21 September 1908	Auvours, France	41·3 miles	(66·5 km)	First major endurance flight. Flew more than 100 times at this location.
3 October 1908	Auvours, France	34·75 miles	(56 km)	In 55 min 37 s.
10 October 1908	Auvours, France	46 miles	(74 km)	In 1 h 9 min 45 s, with M Painleve as passenger.
18 December 1908	Auvours, France	62 miles	(99·8 km)	In 1 h 54 min 53 s. Climbed to 330 ft (100 m) to establish new altitude record.
31 December 1908	Auvours, France	77 miles	(124 km)	In 2 h 20 min 23 s, to win Michelin prize and set up new world record.

Alberto Santos-Dumont

12 November 1906   Made the first flight (accredited) in Europe at Bagatelle, covering 722 ft (220 m) in 21·2 s.

Léon Delagrange

5 November 1907   Covered 500 m at Issy-les-Moulineaux.
11 April 1908   Covered over 2·5 miles (3925 m) at Issy-les-Moulineaux.
23 June 1908   Covered 8·75 miles (14·08 km) in 18 min 30 s at Milan, Italy.
6 September 1908   Covered 15·2 miles (24·4 km) in 29 min 53 s at Issy-les-Moulineaux.
17 September 1908   Flight lasting 30 min 27 s at Issy-les-Moulineaux.

Henry Farman

26 October 1907   Covered 2530 ft (771 m) in 52·6 s.
13 January 1908   Covered a 1 km circuit in 1 min 28 s at Issy-les-Moulineaux.
6 July 1908   Covered 12·4 miles (20 km) in 20 min 20 s.
30 October 1908   Covered 17 miles (27·3 km) in 20 min on cross-country flight from Châlons to Reims.

Glenn H Curtiss

4 July 1908   Covered 5090 ft (1550 m) in 1 min 42 s to win *Scientific American* trophy.

Louis Blériot at Dover, 25 July 1909

the £1000 *Daily Mail* prize for the first Briton to cover a mile (closed circuit) in a British aeroplane – a Short-Wright biplane. Shortly afterwards, he carried a pig in a basket on one of his flights, to demonstrate that the sarcastic expression 'pigs might fly' was thoroughly outdated. He was awarded the Aero Club of Great Britain's Aviator Certificate No 1 on 8 March 1910. Lord Brabazon died in 1969.

The first aeroplane flight of more than one

mile flown in Britain was achieved on 14 May 1909 by Samuel Cody who flew the *British Army Aeroplane No 1* from Laffan's Plain to Danger Hill, Hants – a distance of just over 1 mile – and landed without breaking anything. The Prince of Wales requested a repeat performance during the same afternoon, but Cody, turning to avoid some troops, crashed into an embankment and demolished the tail of his aeroplane.

The first Briton to fly an all-British aeroplane was Alliott Verdon Roe (1877–1958), in his Roe 1 triplane on 13 July 1909 at Lea Marshes, Essex. Lack of funds to build the triplane had forced Roe to construct it from wood instead of light-gauge steel tubing, to cover the wings with paper and to use the same 9 hp JAP engine that had powered his unsuccessful biplane. The 100 ft (30 m) flight that was achieved on the 13th was much improved upon on the 23rd, when he flew 900 ft (275 m) at an average height of 10 ft (3 m).

Conquest of the English Channel. In response to an offer by the *Daily Mail* of a prize of £1000 for the first pilot (of any nationality) to fly an aeroplane across the Channel, **the first attempt** was made by an Englishman, Hubert Latham, flying an Antoinette IV. He took off from

One of Roe's many crashes. The triplane was covered with brown paper, as he could not afford fabric

Sommer and Farman racing at Reims, 1909

Sangatte, near Calais, at 06.42 h on Monday, 19 July 1909, but alighted in the sea after only 6–8 miles (10–13 km) following engine failure which could not be rectified in the air. He was picked up by the French naval vessel *Harpon*. The occasion of this attempt was also the **first instance of wireless telegraphy being used to obtain weather reports,** the first report being transmitted from Sangatte, near Calais, to the Lord Warden Hotel, Dover, at 04.30 h on that morning.

Despite working furiously to get a replacement Antoinette, Latham was beaten by Louis Blériot. The Frenchman took off in his Blériot XI monoplane at 04.41 h, from Les Baraques, on Sunday, 25 July 1909, and landed at 5.17·5 h in the Northfall Meadow by Dover Castle to become **the first man to cross the English Channel in an aeroplane.**

Latham made a second attempt to fly the Channel two days later (on 27 July), taking off at 05.50 h from Cap Blanc Nez. When only 1 mile (1·6 km) from the Dover cliffs, his engine failed and once again he had to alight in the sea.

The first officially recognised aeroplane flight in Russia was made by Van den Schkrouff in a Voisin biplane at Odessa on 25 July 1909.

The first aeroplane flight in Sweden was made by the Frenchman Legagneux at Stockholm in his Voisin biplane on 29 July 1909.

The first woman passenger to fly in an aeroplane in England was Mrs Cody, wife of Samuel, who was taken up by her husband during the last week of July 1909 over Laffan's Plain, Hants, in the *British Army Aeroplane No 1.*

The first passenger to be carried by an aeroplane in Canada was F W 'Casey' Baldwin who was taken aloft on 2 August 1909 at Petawawa, Ontario, in an aeroplane flown by J A D McCurdy.

Left to right: Horace Short, Charles Rolls, Orville Wright, Griffith Brewer and Wilbur Wright in an early Rolls-Royce car. The Shorts built six Wright biplanes at a total price of £8,400 for members of the Aero Club, including Charles Rolls. This was the first manufacture of powered aeroplanes by any UK company

The first aeroplane purchased by the American Government was a Wright biplane, *Miss Columbia*, sold by the Wright brothers on 2 August 1909. The price was $25 000, but a bonus of $5000 was awarded as the specified maximum speed of 40 mile/h (64 km/h) was exceeded. The aircraft was constructed at Dayton, Ohio.

M le Blon's Blériot which is said to have 'crashed in full view of the crowd in a most spectacular manner' at Doncaster, October 1909

The first International Aviation Meeting in the world opened on 22 August 1909 at Reims, and lasted until 29 August 1909. Thirty-eight aeroplanes were entered to participate, although only 23 managed to leave the ground; the meeting also attracted aviators and aeroplane-designers from all over Europe and did much to arouse widespread public interest in flying. The types and numbers of aeroplanes which flew were: Antoinette (3), Blériot XI (2), Blériot XII (1), Blériot XIII (1), Breguet (1), Curtiss (1), Henry Farman (3), REP (1), Voisin (7), Wright (3).

The world's first speed record over 100 km was established by the Englishman, Hubert Latham, during the Reims International Meeting (see above) between 22 and 29 August 1909. Flying an Antoinette (powered by a 50 hp eight-cylinder Antoinette engine) he covered the distance in 1 h 28 min 17 s, at an average speed of 42 mile/h (67 km/h). In so doing he won the second largest prize of the meeting amounting to 42 000 francs. First prize went to Henry Farman who, flying a Gnome-powered Farman, set up new world records for duration and distance in a closed circuit, covering 112½ miles (181·04 km) in 3 h 4 min 56·4 s, winning 63 000 francs.

The first aeroplane flight in the world in which two passengers were carried was made at the Reims International Meeting (see above) on 27 August 1909 by Henry Farman who, in his Gnome-powered Farman biplane, covered a distance of 6·2 miles (10 km) in 10 min 39 s.

The first pilot to be killed flying a powered aeroplane was Eugène Lefebvre, on 7 September 1909; he crashed while flying a new Wright type A at Port Aviation Juvisy. Soon afterwards, on 22 September, Captain Ferber was killed when his Voisin hit a ditch while preparing for take-off.

The first aeroplane flight in Denmark was made by Léon Delagrange in September 1909.

The first Aviation Meeting held in Great Britain was that organised by the Doncaster Aviation Committee on the Doncaster Racecourse between 15 and 23 October 1909. This meeting was not governed by rules laid down by the FAI, nor was it officially recognised by the Aero Club of Great Britain. 12 aeroplanes constituted the field, of which five managed to fly. **The first officially recognised meeting** was held at Squires Gate, Blackpool, between 18 and 23 October 1909, being organised by the Blackpool Corporation and the

Voisin biplane on which Moore-Brabazon learned to fly in France, 1908

Lancashire Aero Club; seven of the dozen participants were coaxed into the air.

The first successful still photographs taken from an aeroplane were by M Meurisse, in December 1909, and showed the flying-fields at Mourmelon and Châlons. The aeroplane was an Antoinette piloted by Latham.

The first aeroplane flight in Ireland is believed to have been carried out by H G Ferguson of Belfast during the winter of 1909–10. The aeroplane was of his own design and manufacture; it resembled a Blériot, and was powered by an eight-cylinder 35 hp air-cooled JAP engine.

Alliott Verdon Roe flying his Roe I triplane at Lea Marshes

The first American monoplane to fly was the Walden III, designed by Dr Henry W Walden and flown on 9 December 1909 at Mineola, Long Island, NY. It was powered by a 22 hp three-cylinder Anzani engine.

The first aeroplane flights in Australia were achieved on 9 December 1909 by Colin Defries, also well known as a motor racing driver. He flew an imported Wright biplane for 1 mile (1·6 km) at a height of 35 ft (11 m) over the Victoria Park Racecourse at Sydney, New South Wales. On the day after his first flight he made a further short flight, this time with a passenger (Mr C S Magennis), **the first to be carried by an aeroplane in Australia.** (It has often been said that Ehrich Weiss, better known as Harry Houdini the escapologist, was the first aeroplane pilot to make a *significant* flight; on 18 March 1910 he flew three times in a Voisin biplane at Digger's Rest, Victoria, achieving a maximum height of over 100 ft (30 m), while his longest flight exceeded 2 miles (3·2 km).

The first military pilot to get the Brevet of the Aéro-Club de France was Lieutenant Camerman on 7 March 1910, receiving Brevet No 33.

The first certificated woman pilot in the world was Mme la Baronne de Laroche, a Frenchwoman, who received her Pilot's Certificate No 36 on 8 March 1910, having qualified on a Voisin biplane. She was killed in 1919 in an aeroplane accident.

The first night flights were made by Emil Aubrun, a Frenchman, on 10 March 1910 flying a Blériot monoplane. Each of the two flights began and ended at Villalugano, Buenos Aires, Argentina, and was about 12·4 miles (20 km) long.

The first take-off from water by an aeroplane was made by Henri Fabre, a Frenchman, in his Gnome-powered monoplane floatplane at Martigues, near Marseille, on 28 March 1910.

The first aeroplane flight in Switzerland was made by Captain Engelhardt at Saint-Moritz on 13 March 1910.

The first recorded night flight in Great Britain was made by Claude Grahame-White during 27/28 April 1910 in his attempt to overhaul Louis Paulhan in the *Daily Mail* £10000 London to Manchester air race. During the course of this race Paulhan (who won) thus made the first London to Manchester flight and was **the first to fly an aeroplane over 100 km (62·14 miles) in a straight line in Great Britain.**

The first aeroplane to be 'forced down' by the action of another was the Henry Farman biplane of Mr A Rawlinson during the Aviation Meeting at Nice, France, in mid April 1910. Mr Rawlinson was flying his new Farman over the sea when the Russian Effimov passed so close above him (also in a Farman) that his downdraught forced the Englishman into the water. The Russian was severely reprimanded for dangerous flying and fined 100 francs.

The first British woman to fly solo in an aeroplane was almost certainly Miss Edith

Mme la Baronne de Laroche

Charles Rolls on his Wright biplane

Maud Cook, who performed various aerial acts under the name of Miss 'Spencer Kavanagh'. She achieved several solo flights on Blériot monoplanes with the Grahame-White Flying School at Pau in the Pyrenees early in 1910. She was also a professional parachute-jumper, known as 'Viola Spencer', and was killed after making a jump from a balloon near Coventry, England, in July 1910.

The first England to France and two-way crossing of the English Channel was accomplished by the Hon C S Rolls (the 'Rolls' of 'Rolls-Royce') flying a French-built Wright biplane on 2 June 1910. He took off from Broadlees, Dover at 18.30·5 h, dropped a letter addressed to the Aero-Club de France near Sangatte at 19.15 h, then flew back to England and made a perfect landing near his starting-rail at 20.06 h. He was thus the **first man to fly from England to France in an aeroplane, the first man to make a non-stop two-way crossing, and the first cross-Channel pilot to land at a pre-arranged spot without damage to his aeroplane.**

The first world record to fall to an Englishman (apart from Henry Farman, and Hubert Latham who established an inaugural record, see page 43) was the duration record taken by Captain Bertram Dickson who, on 6 June 1910, remained airborne for exactly 2 h with a passenger in his Henry Farman biplane at Anjou, France, thereby establishing a new World Endurance Record with one passenger.

The first man to drop missiles from an aeroplane was Glenn Hammond Curtiss on 30 June 1910, when he dropped dummy bombs

Hubert Latham piloting the Antoinette in which he made the first flight to an altitude of 1000 m, 7 January 1910

from a height of 50 ft (15 m) on to the shape of a battleship marked by buoys on Lake Keuka.

The first British pilot to lose his life while flying an aeroplane was the Hon Charles Stewart Rolls (born in London, 27 August 1877, the third son of the first Baron Llangattock), who was killed at the Bournemouth Aviation Week on 12 July 1910 when his French-built Wright biplane suffered a structural failure in flight.

The first flight in Australia by an Australian in an indigenous aeroplane was made by John R Duigan of Mia Mia, Victoria on 16 July 1910 in an aeroplane constructed from photographs of the Wright *Flyer*. On that day Duigan flew only 28 ft (8·5 m), but on 7 October he covered 588 ft (179 m) at a height of about 12 ft (3·65 m).

The first Swedish pilot was Baron Carl Cederström who was granted a Pilot's Certificate at the Blériot Flying School at Pau, France, in 1910. On returning to Sweden he was awarded Certificate No 1 by the Aero Club of Sweden (Svenska Aeronautiska Sallskapet). He was lost, presumed drowned, while flying as a passenger between Stockholm and Finland in July 1918.

The most famous name in early Swedish aviation was probably that of Dr Enoch Thulin, DPhil, who gained his first Pilot's Certificate in France and was subsequently granted Swedish Certificate No 10. Before the First World War he abandoned full-time flying to concentrate upon aeroplane manufacture, establishing his company at Landskrona in 1915. After the United States entered the war, the Thulin Aircraft Works was probably **the largest aircraft manufacturing concern of any neutral country.** Dr Thulin was killed in a flying accident on 14 May 1919.

The first German active duty officer to receive a Pilot's Licence was Leutnant Richard von Tiedemann, a Hussar officer. He first flew solo on 23 July 1910.

The first serving officer of the British Army to be awarded a Pilot's Certificate in England was Captain George William Patrick Dawes who was awarded Certificate No 17 for qualification on a Humber monoplane at Wolverhampton on 26 July 1910.

Captain Dawes died on 17 March 1960 aged 80. He had served in South Africa between 1900

Lieutenant J E Fickel with the rifle which he fired from a Curtiss biplane piloted by Charles F Willard, 20 August 1910

and 1902 when he was awarded the Queen's Medal with three clasps, and the King's Medal with two clasps. He took up flying privately in 1909 and was posted to the RFC on its formation in 1912. He commanded the Corps in the Balkans from 1916 to 1918, during which time he was awarded the DSO and the AFC, was mentioned in despatches seven times, and awarded the Croix de Guerre with three palms, the Serbian Order of the White Eagle, the Order of the Redeemer of Greece and created Officer of the Legion d'Honneur. He served with the Royal Air Force in the Second World War as a Wing Commander, retiring in 1946 with the MBE. He thus was one of the very few officers who served actively in the Boer War and both world wars.

The first mail carried unofficially in an aeroplane in Great Britain was flown by Claude Grahame-White on 10 August 1910 in a Blériot monoplane from Squires Gate, Black-

pool; he did not reach his destination at Southport, having been forced to land by bad weather.

The first Channel crossing with a passenger was by J B Moisant and his mechanic in a Blériot two-seater aeroplane, from Calais to Dover, on 17 August 1910.

The first military firearm to be fired from an aeroplane was a rifle fired by Lieutenant Jacob Earl Fickel, US Army, from a two-seater Curtiss biplane at a target at Sheepshead Bay, New York City, on 20 August 1910.

The first use of radio between an aeroplane and the ground was on 27 August 1910 when James McCurdy, flying a Curtiss, sent and received messages via an H M Horton wireless set at Sheepshead Bay, NY.

The first crossing of the Irish Sea was made by Robert Loraine who, flying a Farman biplane

Eugene Ely leaving the *Pennsylvania* after making earlier the first-ever landing on a ship, 18 January 1911

on 11 September 1910, set off from Holyhead, Anglesey. Although engine failure forced him down in the sea 180 ft (55 m) offshore from the Irish coast near Baily Lighthouse, Howth, he was generally considered to have been the first to accomplish the crossing.

The first flight over the Alps was made by the Peruvian Georges Chavez in a Blériot on 23 September 1910. His flight from Brig to Domodossola, via the Simplon Pass, ended in disaster when he crashed on landing and was killed.

The first air collision in the world occurred on 8 September 1910 between two aeroplanes piloted by brothers named Warchalovski at Wiener-Neustadt, Austria. One of the pilots suffered a broken leg. A passenger on one of the aircraft was the Archduke Leopold-Salvator of Austria.

Former American President Theodore Roosevelt flew three laps of the aerodrome at St Louis in a Wright biplane piloted by A Hoxie in mid October 1910.

Perhaps the greatest feats by a novice pilot

were achieved by T O M (later Sir Thomas) Sopwith in 1910. Having purchased a Howard Wright monoplane, he attempted to perform the first test flight on 22 October, never before having flown as a pilot. It crashed but Sopwith was unhurt. The monoplane was repaired, and he flew it with increasing skill before changing to a Howard Wright biplane. On 21 November, less than a month after his first attempt to fly, he carried out his first-ever taxiing in a biplane before lunch, took off and completed some circuits in the afternoon and qualified for his Pilot's Certificate (No 31) about tea-time. He carried his first passenger the same evening. Five days later he set up new British Distance and Duration Records, which put him in the lead in the competition for the 1910 British Empire Michelin Cup, offered for the longest distance flown in a closed circuit by a British pilot in a British aeroplane. On 18 December 1910 he won the £4000 Baron de Forest prize for the longest distance flown in a straight line into Europe by a British pilot in a British machine during that year. His flight covered 177 miles (285 km) from Eastchurch to Beaumont, Belgium. On the last day

Crissy and Parmelee on the Wright biplane from which they dropped live bombs on 7 January 1911

of the year he regained his lost lead in the Michelin Cup competition by flying 150 miles (241 km), only to be beaten at the very last moment by Cody.

The first aeroplane to take off from a ship was a Curtiss biplane flown by Eugene B Ely from an 83 ft (25 m) platform built over the bows of the American light cruiser USS *Birmingham*, 3750 tons (3810 tonnes), on 14 November 1910. It has often been averred that the vessel was anchored at the time of take-off; this is not correct as it had been proposed to take off as the ship steamed at 20 knots into the wind. In the event, the *Birmingham* had weighed anchor in Hampton Roads, Virginia, but, impatient to take off, Ely gave the signal to release his aircraft at 15.16 h before the ship was under way. With only 57 ft (17 m) of platform ahead of the Curtiss, the aircraft flew off but touched the water and damaged its propeller; the pilot managed to maintain control and landed at Willoughby Spit, 2½ miles (4 km) distant. As

Ely became airborne from the cruiser, the *Birmingham* sent an historic radio message 'Ely's just gone.'

The first explosive bombs dropped by American pilots were those dropped by Lieutenant Myron Sidney Crissy and Philip O Parmelee, from a Wright biplane, during trials on 7 January 1911 at San Francisco, Calif.

The first aeroplane to land on a ship was also a Curtiss biplane flown by Ely, on 18 January 1911, when he landed on a 119 ft 4 in (36 m) long platform constructed over the stern of the American armoured cruiser, USS *Pennsylvania*, 13 680 tons (13 900 tonnes), anchored in San Francisco Bay. It had been intended that the vessel would be under way during the landing, but the Captain considered that there was insufficient sea space to manoeuvre and the *Pennsylvania* remained at anchor. Despite landing downwind the Curtiss rolled to a stop at 11.01 h after a run of only 30 ft (9 m). Captain C F Pond is reputed to have remarked that 'this is the most important landing of a bird since the dove flew back to the Ark'. After lunch Ely successfully took off again from the *Pennsylvania* at 11.58 h and returned to his airfield near San Francisco.

The first aeroplane to perform a premeditated landing on water, taxi and then take off was a Curtiss 'hydroaeroplane' flown by Glenn Curtiss on 26 January 1911. He took off and then landed in San Diego Harbour, turned round and took off again, flying about 1 mile (1·6 km) before coming down near his starting-point. A Curtiss A-1 'hydroaeroplane' was the US Navy's first aeroplane, first flown on 1 July 1911.

The first flight in New Zealand by an aeroplane was made by a Howard Wright type biplane piloted by Vivian C Walsh at Auckland on 5 February 1911. With his brother Leo, Vivian Walsh imported materials from England with which to build the aircraft and installed a 60 hp ENV engine. Vivian Walsh also made **the first seaplane flight in New Zealand** on 1 January 1914.

The first Government (official) air-mail flight in the world was undertaken on 18 February 1911 when the French pilot Henri Pequet flew a Humber biplane from Allahabad to Naini Junction, a distance of about 5 miles (8 km) across the Jumna River, with about 6500 letters. The regular service was established four days later as part of the Universal Postal Exhibition, Allahabad, India, the flights being shared by Captain W G Windham and Pequet. The envelopes of this first air-mail service were franked 'First Aerial Post, UP Exhibition, Allahabad, 1911' and are highly prized among collectors.

The first torpedo drop from an aeroplane was achieved in 1911 by the Italian Capitano Guidoni, flying a Farman biplane. The torpedo weighed 352 lb (160 kg).

Eleven passengers were first carried in an aeroplane on 23 March 1911 by Louis Breguet over a distance of 3·1 miles (5 km) at Douai, France, in a Breguet biplane. **Twelve passengers were first carried in an aeroplane** on 24 March 1911 by Roger Sommer over a distance of 2625 ft (800 m) in a Sommer biplane powered by a 70 hp engine.

The first non-stop flight from London to Paris was made on 12 April 1911 by Pierre Prier in 3 h 56 min, flying a Blériot monoplane powered by a 50 hp Gnome engine. Prier, who was Chief Flying Instructor at the Blériot Flying School, Hendon, took off from Hendon and landed at Issy-les-Moulineaux.

The first recorded carriage of freight by air was a box of Osram lamps carried on 4 July 1911 by a Valkyrie monoplane flown by Horatio Barber from Shoreham to Hove in Sussex, England, on behalf of the General Electric Company who paid £100 for the flight.

The first British woman to be granted a Pilot's Certificate was Mrs Hilda B Hewlett who qualified on a Henry Farman biplane at Brooklands for Certificate No 122 on 29 August 1911. Her son, Sub-Lieutenant F E T Hewlett, RN, was taught to fly by her, and was thus **the first and possibly the only naval airman in the world ever to receive his flying tuition from his mother.** His Certificate, No 156, was gained on 14 November 1911. Young Hewlett was one of the first five officers of the Naval Wing, RFC.

The first official mail to be carried by air in Great Britain was entrusted to the staff pilots of the Grahame-White and Blériot flying schools who commenced carrying the mail between Hendon and Windsor on Saturday, 9 September 1911. The first flight was undertaken on that day by Gustav Hamel in a Blériot monoplane, covering the route in 10 min at a ground speed of over 105 mile/h (169 km/h) with a strong tailwind. The service lasted until 26 September, having been instituted to commemorate the Coronation of HM King George V. The total weight of mail carried between the Hendon flying-field and Royal Farm, Windsor was 1015 lb (460·4 kg).

The first coast-to-coast flight across America was made by Calbraith P Rodgers between 17 September and 5 November 1911. Rodgers, trying to win a $50000 prize offered by William Randolph Hearst, flew from New York to Pasadena in a Wright biplane. Making a series of short flights, he arrived at the destination 19 days outside the specified 30-day limit and so failed to qualify for the prize.

The first official carriage of mail by air in the USA was by Earl L Ovington on 23 September 1911 in a Blériot monoplane from Nassau Boulevard to Mineola, LI, NY, a distance of 6 miles (9·6 km).

Captain Montù with his pilot, G Rossi, about to leave on a reconnaissance flight in North Africa, 31 January 1912. Note the grenades strapped to his chest, ready for dropping on enemy positions

The first gallantry decoration to be 'earned' by a marine aviator was the Distinguished Flying Cross awarded posthumously to Eugene B Ely, who was killed while flying on 14 October 1911. The award of the DFC was made 25 years later in recognition of his outstanding contributions to marine aviation during 1910 and 1911. His sole reward during his life was an award of $500 made by the US Aeronautical Reserve during 1911.

The first Chinese national to receive a Pilot's Certificate was Zee Yee Lee who was awarded Royal Aero Club Certificate No 148 on 17 October 1911, qualifying on a Bristol Boxkite after receiving his training on Salisbury Plain, England. Lee later became Chief Flying Instructor at the Military Flying School at Nanyuen, Peking. He was followed by Prince Tsai Tao, Wee Gee, Colonel Tsing, Lieutenant Poa and Lieutenant Yoa, at least two of whom gained Certificates in the USA.

The first time an aeroplane was used in war was on 22 October 1911 when an Italian Blériot, piloted by Capitano Piazza, made a reconnaissance flight from Tripoli to Azizia to view the Turkish positions.

The first officer of the Royal Navy to take off from a ship in an aeroplane was Lieutenant Charles Rumney Samson who is said to have made a secret flight in a Short S.27 from a platform on the bows of the British battleship, HMS *Africa*, 17 500 tons (17 780 tonnes), moored in Sheerness Harbour during December 1911. His first officially recorded take-off was from HMS *Africa* at 14.20 h on 10 January 1912, flying a modified Short biplane. Commander Samson was appointed Officer Commanding the Naval Wing of the Royal Flying Corps in October 1912.

The first seaplane competition was held at Monaco in March 1912. Seven pilots attended (Fischer, Renaux, Paulhan, Robinson, Caudron, Benoit, Rugère), the winner being Fischer on a Henry Farman biplane.

The first instance of a Government ordering the grounding of a specific type of aircraft occurred in March 1912 when the French Government ordered all Blériot monoplanes of the French Army to be prohibited from flying until they had been rebuilt so that their wings were braced to withstand a degree of negative-G. Five distinguished French pilots had been killed following the collapse of the Blériot's wings, but the ban was short-lived and the aircraft were flying again within a fortnight. The weakness was spotlighted by Louis Blériot himself who, despite the likely loss of prestige, published a short report explaining the weakness in his own aeroplanes. There is no doubt that his frankness increased – rather than detracted from – his very high standing in aviation circles.

The first pilot in the world to take off in an aeroplane from a ship under way was Commander Samson, who took off in a Short pusher biplane amphibian from the forecastle of the battleship HMS *Hibernia* while it steamed at 10·5 knots off Portland during the Naval Review of May 1912. At the conclusion of the Review, Commander Samson was one of the officers commanded to dine with HM King George V on board the *Victoria and Albert*.

The first single-seat scout aeroplane was the Farnborough BS1 of 1912 which was designed mainly by Geoffrey de Havilland.

Another Farnborough aeroplane, the BE1, made **the first successful artillery-spotting flight** over Salisbury Plain in 1912.

The first American woman to receive her Pilot's Certificate was Harriet Quimby. She was also the **first woman to pilot an aeroplane across the English Channel,** on 16 April 1912. Taking off from Deal, in a Blériot monoplane, she landed at Cape Gris-Nez less than an hour later.

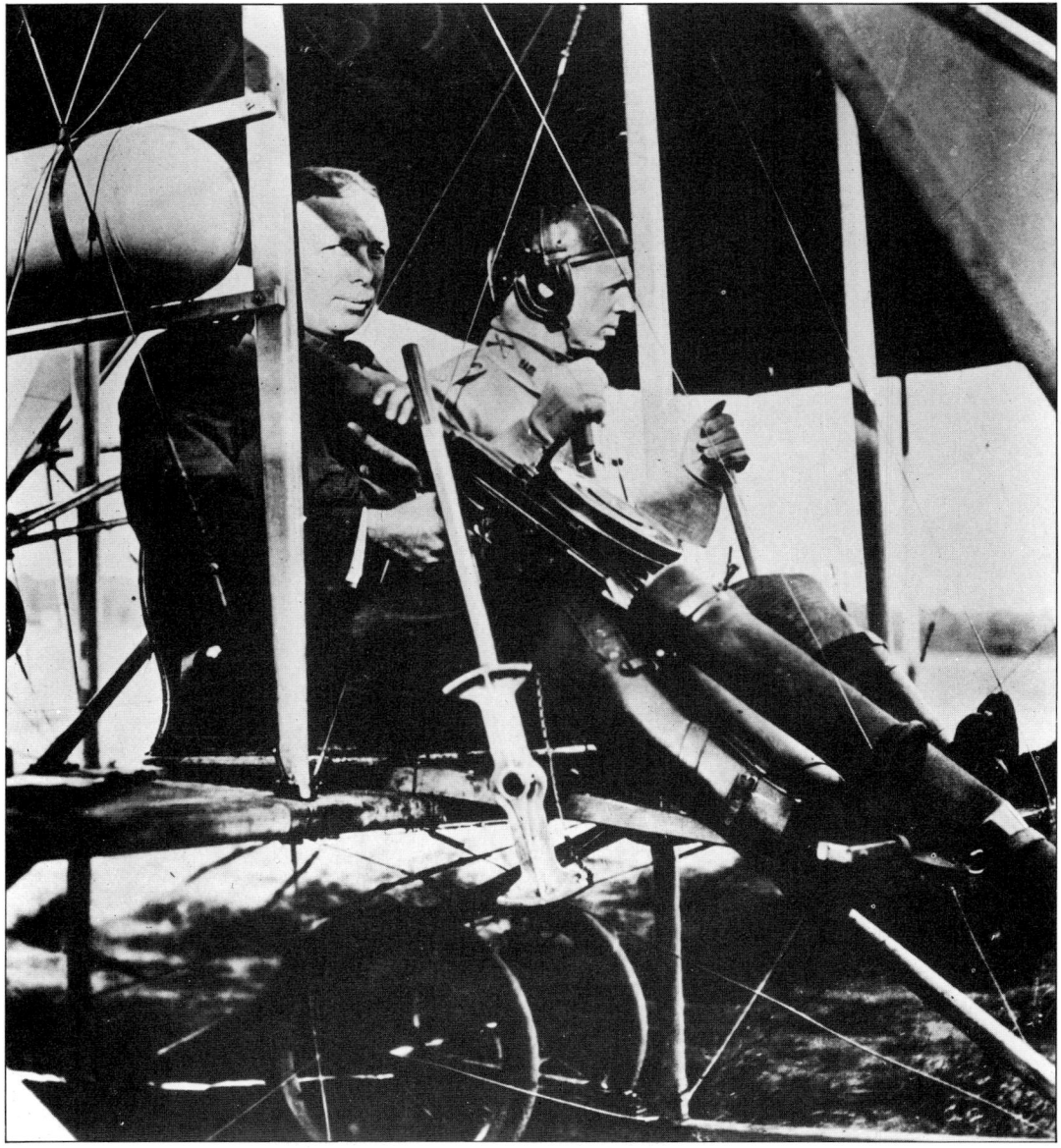

Charles de Forest Chandler with the first machine-gun fired from an aeroplane, a Wright Type B, in June 1912

The first flight from Paris to Berlin was achieved by Edmond Audemars of Switzerland who flew a Blériot monoplane from the French capital to the German capital via Bochum in Westphalia during the spring of 1912.

The first all-metal aeroplane to fly was the Tubavion monoplane built by the Frenchmen Ponche and Primard in 1912. A fatal accident brought its tests to a halt.

The first flight in Norway by a Norwegian took place on 1 June 1912 when Lieutenant Hans E Dons, a submarine officer, flew a German Start across Oslo Fjord from Horten to Frederikstad, a distance of 37 miles (60 km). As a result of this achievement the Norwegian Storting (Government) voted the sum of £1250 ($2160) to send four officers to Paris to learn to fly. Within three months (in August) one of these officers had established a Scandinavian distance record.

The first American woman to be killed in an aeroplane accident was Julie Clark of Denver, Colorado, whose Curtiss biplane struck a tree on 17 June 1912 at Springfield, Illinois, and turned over. She had qualified for her Pilot's Certificate on 19 May 1912.

The first American aeroplane armed with a machine-gun was a Wright biplane flown by Lieutenant Thomas de Witt Milling at College Park, Md, on 2 June 1912. The gunner, who was armed with a Lewis gun, was Charles de Forest Chandler of the US Army Signal Corps.

The first crossing of the English Channel by an aeroplane with a pilot and two passengers was made on 4 August 1912 by W B Rhodes Moorhouse (later, as a Lieutenant in the Royal Flying Corps, the first British airman to be awarded the Victoria Cross on 26 April 1915) who, accompanied by his wife and a friend, flew a Breguet tractor biplane from Douai, France, via Boulogne and Dungeness, to Bethersden, near Ashford, Kent, where they crashed in bad weather. Nobody was hurt.

The first man to fly underneath all the Thames bridges in London between Tower Bridge and Westminster was F K McClean who, flying a Short pusher biplane from Harty Ferry, Isle of Sheppey, in mid August 1912, passed between the upper and lower spans of Tower Bridge, and then underflew all the remaining bridges to Westminster where he landed on the river. No regulations forbade this escapade, but the police instructed McClean to taxi all the way back to Shadwell Basin before mooring! On the return trip the aeroplane side-slipped soon after take-off and damaged one of the floats after hitting a barge. The machine was then towed into Shadwell Dock and dismantled for the return by road to Eastchurch.

The first officer of the Royal Flying Corps Reserves to be killed while engaged on military flying duties was Second-Lieutenant E Hotchkiss (the Bristol Company's Chief Flying Instructor at Brooklands) who, with Lieutenant C A Bettington, was killed on 10 September 1912 when their Bristol monoplane crashed on a flight from Salisbury Plain. The aircraft suffered a structural failure, after which the wing fabric started to tear away and the aircraft crashed near Oxford. Within three weeks the flying of monoplanes by the Military Wing was banned by Colonel Seely, Secretary of State for

FORMATION OF THE FIRST BRITISH MILITARY AEROPLANE SQUADRONS

(The Royal Flying Corps came into being officially on 13 May 1912.)

Squadron	Date	Remarks
No 1 (Airship and Kite)	13 May 1912	Formed out of No 1 Airship Company, Air Battalion.
No 2 (Aeroplane)	13 May 1912	Formed from scratch.
No 3 (Aeroplane)	13 May 1912	Formed out of No 2 Aeroplane Company, Air Battalion.
No 4 (Aeroplane)	16 May 1912	Formed from scratch.
No 5 (Aeroplane)	26 July 1913	Formed from scratch at Farnborough.
No 6 (Aeroplane)	31 January 1914	Formed from scratch at Farnborough.
No 7 (Aeroplane)	May 1914	Formed from scratch.
No 8 (Aeroplane)	May 1914	Formed out of No 1 Airship and Kite Squadron, RFC.

Igor Sikorsky with a model of *Le Grand*, the first four-engined aeroplane. Photo taken in the early 1950s

War, and although the ban was to last no more than five months it gave rise to an extraordinary prejudice against monoplanes in British military flying circles that was to persist for more than 20 years. (It is usually recorded that Hotchkiss's Bristol crashed on Port Meadow, Oxford, but the memorial tablet confirms that in fact the accident occurred half a mile west of Godstow on the right bank of the River Thames, just north of Port Meadow.)

The first aeroplane to be successfully catapult-launched from a boat was the Curtiss A-1 floatplane, piloted by Lieutenant T Ellyson, on 12 November 1912. The operation was performed from an anchored barge, at the Washington Navy Yard, using a compressed-air launcher invented by Captain W I Chambers.

The number of Pilots' Certificates which had been awarded in the world by the end

of 1912 was 2480, though the number of actual pilots was slightly smaller as some had been awarded certificates in more than one country. One or two others had received certificates in countries which were not members of the Fédération Aéronautique Internationale. The massive superiority of France at this time is evident:

1	France	966	10	Holland	26
2	Great Britain	382	11	Argentine	
3	Germany	345		Republic	15
4	United States of			Spain	15
	America	193	13	Sweden	10
5	Italy	186	14	Denmark	8
6	Russia	162	15	Hungary	7
7	Austria	84	16	Norway	5
8	Belgium	58	17	Egypt	1
9	Switzerland	27			

Total   2490

Two British pilots had been killed during 1910; seven aeroplane occupants were killed in 1911, 17 in 1912 and 15 in 1913. By the outbreak of the First World War 78 British subjects had lost their lives in aeroplanes at home or abroad.

The first military aircraft acquired by China were six 80 hp and six 50 hp Caudrons ordered from France in March 1913.

The first night flight by a British military aircraft took place either on the night of 15/16 or 16/17 April 1913. Lieutenant R Cholmondeley, of No 3 Squadron, Military Wing, RFC, flew a Maurice Farman biplane from Larkhill to Upavon and back by moonlight.

The first gyroscopic automatic stabiliser was successfully demonstrated by the Americans, Lawrence B Sperry and Lieutenant Patrick Nelson Lynch Bellinger, in a Curtiss F flying-boat in 1913. The aircraft was longitudinally and laterally stabilised.

The first four-engined aeroplane to fly was the *Le Grand* ('The Great One') biplane designed, and first flown on 13 May 1913 at St Petersburg, by Igor Sikorsky, then Head of the Aeronautical Department of the Russian Baltic Railway Car Factory at Petrograd. It had a wing span of over 92 ft (28 m) and was powered by four 100 hp Argus engines. From *Le Grand* was evolved the Ilya Mourometz, which became the **first four-engined bomber to see active service.**

The first air crossing of the Mediterranean was achieved on 23 September 1913 by a Morane-Saulnier monoplane piloted by Roland Garros, who flew 453 miles (700 km) from Saint-Raphaël, France, to Bizerte, Tunisia, in 7 h 53 min.

The first flight from France to Egypt was accomplished by Jules Védrines in a Blériot powered by an 80 hp Gnome engine, between 29 November and 29 December 1913. Setting out from Nancy, France, his route was via Würzburg, Prague, Vienna, Belgrade, Sofia, Constantinople, Tripoli (Syria), Jaffa and Cairo.

The first scheduled airline using aeroplanes was the Benoist Company airline which started its operations in January 1914, flying between St Petersburg and Tampa, Florida. The aircraft was a Benoist flying-boat piloted by A Jannus. The operation lasted four months.

The first French airmen to be killed on active service were Captain Hervé and his observer, named Roëland. During the colonial campaign in Morocco early in 1914 they made a forced landing in the desert and were killed by local Arabs.

The first military operations involving the use of American aeroplanes were those against Vera Cruz, Mexico, in April 1914 when five Curtiss AB flying-boats were carried to the port on board the battleship USS *Mississippi* and the cruiser USS *Birmingham*. The first such military flight was undertaken by Lieutenant (Jg) P N L Bellinger who took off in the Curtiss AB-3 flying-boat on 25 April in order to search for mines in the harbour. The ABs came under rifle fire, which caused damage but no loss of aircraft or pilots.

The first passenger to be carried from one city to another in Canada by air was flown in the Curtiss flying-boat *Sunfish* from Toronto to Hamilton and back by Theodore Macaulay on 15 May 1914.

The first Air Service of the US Army was established on 18 July 1914 when an aviation section was formed as part of the Signal Corps with a 'paper' strength of 60 officers and 260 men. The entire equipment amounted to six aeroplanes.

The first standard naval torpedo dropped by a naval airman in a naval aircraft was a 14 in (35·6 cm) torpedo weighing 810 lb (367 kg), dropped by a Short seaplane flown by Squadron Commander Arthur Longmore, RN (Royal Aero Club Pilot's Certificate No 72), on 28 July 1914.

The first flight across the North Sea by an aeroplane was achieved by the Norwegian pilot Tryggve Gran flying a Blériot monoplane on 30 July 1914.

The first British airline company to be registered was Aircraft Transport and Travel Ltd. It was registered in London on 5 October 1916 by George Holt Thomas.

The first scheduled regular international air-mail service in the world was inaugurated between Vienna and Kiev, via Kraków, Lvóv and Proskurov on 11 March 1918. The service was principally for military mails and was operated with Hansa-Brandenburg C I biplanes, continuing until November 1918.

The first air crossing of the Andes was achieved by the Argentine army pilot Teniente Luis C Candelaria flying a Morane-Saulnier parasol monoplane on 13 April 1918 from Zapala, Argentina, to Cunco, Chile, a distance of approximately 124 miles (200 km). The maximum altitude was about 13 000 ft (4000 m). Candelaria had attended the fifth military flying course at El Palomar which commenced in September 1916.

The first experimental air-mail service in the USA was flown by War Department Curtiss JN-4 aircraft on 15 May 1918 between Washington, DC, Philadelphia, Pa, and New York City. Lieutenant Torrey H Webb was the first pilot.

The first official air-mail flight in Canada was flown on 24 June 1918 in a Curtiss JN-4 from Montreal to Toronto by Captain Brian A Peck, RAF, accompanied by Corporal Mathers.

The first variable-incidence variable-geometry aeroplane in the world was the Swedish Pålson Type 1 single-seat sporting aircraft of 1918–19. It is said that the aircraft featured a system of cranks to alter the position of the biplane's top wing as well as its angle of incidence as a means of achieving optimum lift/drag in cruising flight. It is not known what success attended flight trials (if any).

Line-up of RFC aeroplanes at the Central Flying School, Upavon, 1914. Types illustrated are an Avro 500, a Blériot XI and several Henry Farman biplanes

SECTION 4

THE FIRST WORLD WAR

No sooner had hot-air and hydrogen balloons been invented, than ways and means were sought to adapt them for military use. In 1794 a man-carrying balloon was used by the French Republican Army for observation duties; five years later it was suggested that the French and Spanish fleets could be destroyed by a huge number of attacking balloons, from which a variety of weapons could be hurled upon the helpless ships. In the latter half of the 19th century, balloon corps were formed in several armies for observation duties. Although the balloon was used during the First World War for similar tasks, it was with the powered aeroplane that aerial warfare really began in earnest.

Several major events occurred in the first eight years of practical powered flight which should have had more impact than they did on the immediate pre-war years. In 1908 the US Army had tested and bought a Wright biplane; in November 1910 Eugene Ely proved the aircraft carrier a practical idea; and from October 1911 Italy used aeroplanes to observe and, later, bomb Turkish forces. The first bombing raid was made from a bird-like Etrich Taube monoplane on 1 November. No less significant were the observation, photographic and bombing raids carried out by Spanish aircraft in Morocco in 1913.

Even after these early developments, little more was done before 1914 to develop the aeroplane into an efficient attacking machine; although it is known that German agents kept a close eye on developments in military flying in Britain from 1912. Taube monoplanes were used again, by the German forces, during the first stages of the war, for reconnaissance and the first light bombing attack on Paris. Air combat was still in the future; for although guns had been fired from several types of aeroplane since August 1910, armed aircraft were not seen in service until late 1914.

When war was declared in August of that year, the Imperial German Army and Navy air services had about 280 assorted aircraft and 9 airships. Austria-Hungary had 36; Britain had about 180 aircraft, Belgium had 24, and France had 160 aircraft and 15 airships. These were used initially for reconnaissance, artillery spotting and light bombing duties, and soon became the eyes of the armies. This pointed, inevitably, to the need for armed aircraft or fighter-scouts to destroy opposing machines before vital information could reach enemy land forces.

Arming an aeroplane was far from easy until the invention of interrupter gear, which allowed a machine-gun to fire between the rotating propeller blades. The Germans were first to fit such an arrangement, on the Fokker Eindecker, and gained almost complete control of the skies over the Western Front during the Winter of 1915–16. The British answer was to build pusher fighters, or aircraft with rear-mounted engines and propellers, so leaving a clear forward field of fire. However, the future lay with front-engined aircraft, and the Allies had to invent their own interrupter gear.

As the war progressed, fighters got faster and bombers got heavier, although no British or German production aircraft reached the 131 mph claimed by the British S.E.4 built at the Royal Aircraft Establishment in pre-war days. Superiority was held first by one side, then the other, as each air force gained that slight advantage in speed or manoeuvrability over its enemies. Despite this, there was often great respect for opposing pilots, and some aces preferred to force down or damage an enemy machine rather than to set it on fire and so condemn the pilot to endure a most painful and terrifying death at a time before parachutes were carried.

After the war had been won by the Allies in 1918, hostilities continued in Russia during 1919. However, the Red Russian forces gradually gained control, and established a permanent Communist government. Now the First World War was really over. The huge aircraft industry of Britain, in particular, had already been axed overnight. As the armament factories closed, massive unemployment set in. Many disillusioned soldiers and airmen returning from the Fronts asked 'Where is the promised country fit for heroes to live in?'

A brief list of the declarations of war by the major powers, and other dates, which formed the basis for the First World War.

On 1 August 1914 Germany declared war on Russia.

On 2 August 1914 Germany invaded Luxembourg.

On 3 August 1914 Germany declared war on France.

On 3/4 August 1914 Germany invaded Belgium.

On 4 August 1914 Britain declared war on Germany.

On 5 August 1914 Austria–Hungary declared war on Russia.

On 6 August 1914 Serbia declared war on Germany.

On 10 August 1914 France declared war on Austria–Hungary.

On 12 August 1914 Britain declared war on Austria–Hungary.

On 23 August 1914 Japan declared war on Germany.

On 25 August 1914 Austria–Hungary declared war on Japan.

On 28 August 1914 Austria–Hungary declared war on Belgium.

On 31 October 1914 Russia declared war on Turkey.

On 5 November 1914 Britain declared war on Turkey.

On 24 May 1915 Italy declared war on Austria–Hungary.

On 27 August 1916 Italy declared war on Germany.

On 27 August 1916 Romania declared war on Austria–Hungary.

On 28 August 1916 Germany declared war on Romania.

On 30 August 1916 Turkey declared war on Romania.

On 1 September 1916 Bulgaria declared war on Romania.

On 6 April 1917 the United States of America declared war on Germany.

On 7 December 1917 the United States of America declared war on Austria–Hungary.

The first British airmen killed on active service were Second-Lieutenant R B Skene and a mechanic, R K Barlow, of No 3 Squadron, RFC, on 12 August 1914. Flying from Netheravon to Dover to form up for the Channel crossing, their aircraft, a Blériot two-seater of 'C' Flight, landed because of engine trouble. Shortly after taking off again, the aircraft crashed into trees and both occupants were killed.

The first British squadrons to fly over the English Channel to France after the outbreak of war were numbers 2, 3, 4 and 5, equipped with BE2s, Blériots and Farman biplanes; BE2s and Farmans; and BE8s and Avro 504s respectively, starting on 13 August 1914. Farmans of No 4 Squadron were later the **first British armed aircraft to be flown in action.**

The first German Air Service pilot to be killed on active service was Oberleutnant Reinhold Jahnow. He was fatally injured in a crash at Malmédy, Belgium, on 12 August 1914. He was holder of German Pilot's Licence No 80, and a veteran of several reconnaissance flights for the Turks during the Balkan campaign of 1912.

The first bombing attack of the war was made by Lieutenant Césari and Corporal Pindhommeau of the French Air Force, against Zeppelin sheds at Metz-Frescaty, on 14 August 1914.

The most widely used aeroplane type in military service on the outbreak of war was the Etrich Taube. Designed in Austria–Hungary in 1910, it was a 'bird-winged' monoplane powered by a single engine of 85–120 hp. Maximum speed was 72 mph (116 km/h). In August 1914, about half of all the aircraft in German service were of this type and others were operated by the Austro–Hungarian air service. Intended mainly for reconnaissance and training duties, Taubes were often used for dropping light bombs. Germany alone licence-built some 500, examples of which remained in service until 1916.

The first British reconnaissance flight over German territory was carried out by Lieutenant G Mappleback and Captain P Joubert de la Ferté of No 4 Squadron, RFC, flying a BE2b and a Blériot monoplane respectively. The flight took place on 19 August 1914.

The first RFC aeroplane to be brought down in action was an Avro 504 of No 5 Squadron, piloted by Lieutenant V Waterfall, on 22 August 1914. The aircraft was shot down by rifle fire from troops in Belgium.

The first British military aircraft insignia consisted of Union Jacks painted in rectangular and shield-shape forms on RFC aircraft. This was necessitated by the fact that RFC aircraft had been fired on by French and British ground-troops, who mistook them for German types. RNAS aircraft were instructed to bear the Union Jack on 26 October 1914. **The roundel was adopted by the RFC from 11 December 1914,** following the French example. On 11 December 1914, the RNAS adopted a roundel for the wings only, consisting initially of a red outer circle and a white centre.

The first bombs to be dropped upon a capital city from an aircraft fell on Paris, on 30 August 1914. The pilot of the German Taube aeroplane has been variously identified as Leutnant Franz von Hiddeson and Leutnant Ferdinand von Hiddessen. (The confusion between names was caused by the message that was dropped with the bombs reading 'The German Army is at your gates – you can do nothing but surrender, Leut von Heidssen.') Five bombs were dropped, killing one woman and injuring two other persons.

The first enemy aircraft forced down in combat by British aircraft was a German two-seater forced to land on 25 August 1914 by three aircraft of No 2 Squadron, RFC. The pilot who finally forced it down was Lieutenant H D Harvey-Kelly who had been **the first RFC pilot to land in France** on 13 August in a BE2a (No 347).

The Royal Aircraft Factory BE2 was produced in large numbers for the RFC and RAF mainly as two-seat reconnaissance and light bombing aircraft, although the BE2c, d and e versions were armed with two machine-guns and served abroad and with home defence units as night fighters against raiding airships. BE2s were among the first aircraft to be sent to France, and over 3240 were eventually produced, of several versions. Designed to have good inherent stability, to allow easy flying while observing or photographing the land below, this very characteristic led to heavy losses as BE2s could not easily be manoeuvred when attacked. Furthermore, production of the series continued after development of newer combat aircraft had made the type obsolete. Powered by engines ranging from 70–90 hp, the maximum speed was between 70 and 72 mph (113–116 km/h).

The first aeroplane to be destroyed by ramming was an Austrian two-seater flown by Leutnant Baron von Rosenthal, rammed over Galicia on 26 August 1914 by Staff Captain Petr Nikolaevich Nesterov of the Imperial Russian

Crew of the Voisin of Escadrille VB24 who achieved the first-ever victory in air combat

XI Corps Air Squadron, who was flying an unarmed Morane Type M monoplane scout. Both pilots were killed. Nesterov (remembered also as the **first pilot to loop the loop**) was the Imperial Air Service's **first battle casualty**.

The first great land battle in which victory was generally attributed to aerial reconnaissance was the battle of Tannenberg, where 125000 Russian soldiers and 500 guns were captured by German forces in late August 1914.

The first RNAS Squadron to fly to France after the start of the war was the Eastchurch Squadron, led by Wing Commander C R Samson (the first British pilot to take off in an aeroplane from a ship, etc.). Arriving at Ostend on 27 August 1914, its equipment included two Sopwith Tabloids, three BEs, two Blériots, one Short seaplane, one Bristol biplane and one Farman biplane. The only armed aircraft attached to the Squadron was the Astra-Torres airship No 3.

The first air operations undertaken by airmen of the Royal Navy during the First World War were reconnaissance flights by Eastchurch Squadron commanded by Wing Commander Charles Samson in support of a Brigade of Royal Marines on the Belgian coast in August 1914.

Fléchettes were first dropped from aeroplanes of No 3 Squadron RFC in the autumn of 1914. These were steel darts about 5 in (12·7 cm) in length which were carried in containers. Over the target, some 250 fléchettes were dropped from each container on to enemy ground concentrations. Casualties or damage were rare in such attacks.

The first British air raid on Germany was by four aircraft of the Eastchurch RNAS Squadron. On 22 September 1914 two aircraft took off from Antwerp to attack the airship sheds at Düsseldorf, two to attack the airship sheds at Cologne. Only the aircraft flown by Flight-Lieutenant Collet found the target – the sheds at Düsseldorf – and his three 20 lb (9 kg) Hales bombs, while probably on target, failed to explode. All aircraft returned safely.

The Iron Cross was first seen painted on German aircraft in late September 1914.

The first aeroplane in the world to be shot down and destroyed by another was a German two-seater, possibly an Aviatik, shot down

at Jonchery, near Reims on 5 October 1914 by Sergent Joseph Frantz and Caporal Quénault in a Voisin pusher of Escadrille VB24. The weapon used is believed to have been a Hotchkiss machine-gun.

The first successful British air raid on Germany took place on 8 October 1914. Squadron Commander D A Spenser Grey and Flight-Lieutenant R L G Marix of the Eastchurch RNAS Squadron flew from Antwerp in Sopwith Tabloids (Nos 167 and 168) to attack airship sheds at Düsseldorf and Cologne with 20 lb (9 kg) Hales bombs. Grey failed to find the target, bombed Cologne Railway Station and returned to Antwerp. Marix reached his target at Düsseldorf, bombed the shed from 600 ft (200 m) and destroyed it and Zeppelin Z.IX inside. His aircraft was damaged by gun-fire, and he eventually crash-landed 20 miles (30 km) from Antwerp, returning to the city on a bicycle borrowed from a peasant.

The Sopwith Tabloid was the first single-seat scout to enter production for military service in the world. Designed before the war, a Tabloid won the 1914 Schneider Trophy contest, and later became standard equipment of the early RNAS. Examples serving with the Eastchurch Squadron were armed with a wing-mounted machine-gun from February 1915.

The first operational seaplane unit of the Imperial German Navy was formed on 4 December 1914, moving to its base at Zeebrugge two days later.

The Avro 504 in which Squadron Commander E F Briggs attacked the Zeppelin sheds at Friedrichshafen on 21 November 1914. This was the first-ever strategic bombing raid by a formation of aircraft, as Briggs was accompanied by Flight Commander J T Babington and Flight Lieutenant S V Sippé, also on 504s of the Royal Naval Air Service. The aircraft flew from Belfort in France, each carrying four 20 lb bombs, with which a Zeppelin was damaged in its shed and the gasworks destroyed. Briggs was shot down and taken prisoner.

BE2a *of the Royal Naval Air Service, Eastchurch Squadron, flown by Squadron Commander Charles R Samson in France and Belgium during the first two years of the First World War. Samson formed an affection for No 50, and it accompanied him to Tenedos when he was posted to command naval aircraft in the Dardanelles campaign; in the course of the campaign he continued to fly the BE2a on reconnaissance, artillery spotting and bombing missions against the Turkish forces. Details of Samson's career may be found in the text.*

Nieuport 17 B1566, *flown during the early summer of 1917 by Captain William A Bishop, DSO, MC, of No 60 Squadron RFC. Bishop, whose career is described in fuller detail in the text, is thought to have achieved nearly twenty of his seventy-two confirmed victories while flying B1566.*

Sopwith Triplane N5492 Black Maria, *flown during the spring and early summer of 1917 by Flight Sub-Lieutenant Raymond Collishaw as commander of 'B' Flight No 10 Suadron, RNAS. During twenty-seven days of June 1917 Collishaw flew this aircraft to victory over sixteen enemy machines, including the Albatros D III of Jasta II flown by the ace, Leutnant Karl Allmenröder. N5492 was eventually shot down in July while being flown in combat by another pilot. Details of Collishaw's career may be found in the text.*

SE5a B4863, *one of the aircraft flown during the summer and autumn of 1941 by Captain James T B McCudden, MC, MM, as commander of 'B' Flight, No 56 Squadron, RFC. Details of McCudden's career may be found in the text. During the winter of 1917/18 he is known to have flown SE5a B4891, fitted with the red-painted propeller spinner from an LVG C V two-seater which he shot down on 30 November.*

Sopwith 7F1 Snipe E8102, *flown on 27 October 1918 by Major William G Barker, DSO, MC, attached to No 201 Squadron, RAF. Flying alone on the early morning of 27 October, Barker had shot down a German two-seater when he was attacked and wounded in the right thigh by a Fokker D VII. Barker's aircraft lost height in a spin; in the course of the next few minutes he passed through successive layers of a large German formation, being attacked on four separate occasions by groups of at least a dozen Fokker scouts. Before he finally managed to bring his damaged Snipe down for a successful forced landing, Barker had been wounded twice more (in the left thigh and the left elbow); had lost consciousness twice, and twice recovered and regained control of his aircraft; and had shot down three more enemy aircraft. This epic engagement led to the award of the Victoria Cross.*

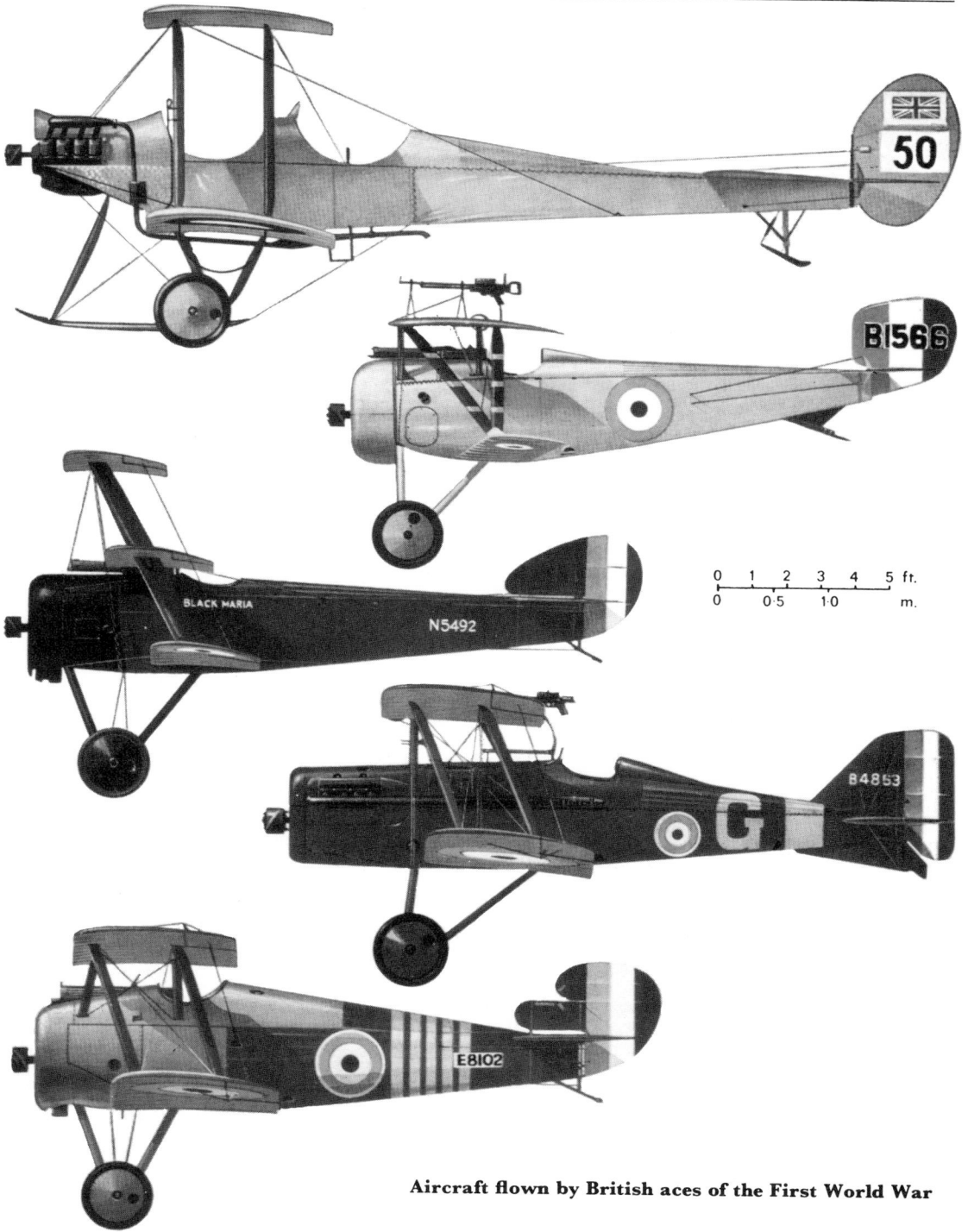

Aircraft flown by British aces of the First World War

The first aeroplane raid on Great Britain, by one aircraft, took place on 21 December 1914. Two bombs fell in the sea near Admiralty Pier at Dover.

The first bomb dropped by an enemy aircraft on British soil, and the second aeroplane raid on Great Britain, again by one aircraft, took place on 24 December 1914. One bomb exploded near Dover Castle.

The first four-engined bomber to become operational was the Russian Sikorsky Ilya Mourometz, of which 73 were delivered to the Czar's Squadron of Flying Ships. Between 1915 and 1917, those not used for training made about 400 bombing raids, the first against a Polish target on 15 February 1915. Only one was shot down by another aircraft. Powered by four 125–220 hp engines, the Ilya Mourometz could fly at 60–80 mph (97–129 km/h) and could carry up to 1500 lb (680 kg) of bombs. It carried up to 16 crew members and, as with the German R-Type bombers of 1918, routine servicing and minor repairs could be carried out in flight.

The first airship raid on Great Britain was carried out on 19 January 1915 by three German Navy Zeppelins, L3, L4 and L6. They took off from Fuhlsbüttel and Nordholz. L6 was forced to return through engine trouble but L3 and L4 arrived over the Norfolk coast at about 20.00 h; nine bombs were dropped in the Great Yarmouth area at 20.25 h by L3, killing two persons and wounding three others. Meanwhile L4 had gone north-west towards Bacton and dropped incendiary bombs on Sheringham, Thornham and Brancaster as well as a high-explosive bomb on Hunstanton wireless-station. Following that, it dropped bombs on Heacham, Snettisham and King's Lynn, where seven high-explosive bombs were dropped and an incendiary, killing

Sikorsky Ilya Mourometz bomber of the Czar's Squadron of Flying Ships

A 920 lb bomb dropped from an Ilya Mourometz bomber. In a flat cap second from left of bomb is Igor Sikorsky. In a similar position to the right is General Michael Shidlowsky, who commanded the Squadron of Flying Ships

two people and injuring 13. The two airships were both wrecked on the coast of Jutland on 17 February 1915 after running into a gale on their homeward journey after trying to spot the British Fleet. Altogether, **some 88 Zeppelins were constructed during the war.**

The first use of aeroplanes in military operations in South America was in February 1915, by the Brazilian Army in the State of Santa Catarina.

The first aeroplane to be designed and built for aerial fighting was the Vickers FB5 Gunbus. Armed with one forward-firing machine-gun, which was operated by a second crew member, FB5s started reaching units in France in February 1915, and the first FB5 fighter squadron was formed in July. Powered by a 100 hp Gnome rotary engine in 'pusher' configuration, the FB5 had a maximum speed of 70 mph (113 km/h).

Turkey's attack on the Suez Canal, which

began on 3 February 1915, was repelled partly because of an aerial reconnaissance carried out on 23 January which located the advancing Turkish troops as they approached the area.

The first naval vessel fully converted for aircraft duties, while still under construction was HMS *Ark Royal,* and as such was the first ship in the world to be completed as an aircraft (seaplane) carrier. Launched in 1914 *Ark Royal* became the first aircraft carrier to operate aeroplanes against the enemy in Europe (the *Wakamiya Maru* had launched seaplanes against the Germans in the Far East by this time) when, arriving at the entrance to the Dardanelles on 17 February 1915, one of her seaplanes was sent on reconnaissance against the Turks.

The first British bombing raid in direct tactical support of a ground operation occurred on 10 March 1915, comprising attacks on railways bringing up German reinforcements in the Menin and Courtrai areas (Second Wing)

Morane-Saulnier monoplane fitted with deflector plates to prevent bullets from the forward-firing machine-gun hitting its propeller (Imperial War Museum)

and the railway stations at Lille, Douai and Don (bombed by the Third Wing), during the Neuve Chapelle offensive. The Divisional Headquarters at Fournes was also bombed by three aircraft of No 3 Squadron piloted by Captain E L Conran, Lieutenant W C Birch and Lieutenant D R Hanlon.

The first single-seat fighter to destroy an enemy aircraft using a machine-gun that fired through the propeller disc was a French Morane-Saulnier Type M piloted by Roland Garros. Having first fitted deflector plates to the propeller to prevent the bullets from hitting the rotating blades, Garros claimed his first victory using this method on 1 April 1915. On 19 April Garros had to make an emergency landing behind German lines and his aircraft, along with its secret, was captured.

The first air Victoria Cross was awarded post-humously to Lieutenant W B Rhodes Moor-house, pilot of a BE2 of No 2 Squadron, RFC, for gallantry in a low-level bombing attack on Courtrai railway station on 26 April 1915.

The aircraft sent by the Allies to take part in the campaign against German controlled South West Africa were three Henry Farman and two Royal Aircraft Factory BE2c biplanes. These arrived on the last day of April 1915 at Walvis Bay and were used in conjunction with the Union Expeditionary Force. The campaign was successfully concluded on 9 July 1915. The Allied unit comprised mostly South African personnel.

The first fighter aircraft to be fitted with a synchronized machine-gun, firing forward between the propeller blades, was the German-built Fokker Eindecker. About 425 'E' series monoplanes were built. None flew faster than

87 mph (140 km/h), but they caused such havoc in attacks on Allied aircraft that their activities for ten months in 1915–16 are remembered as the 'Fokker Scourge'. The inherently stable BE2cs of the RAF suffered particularly heavy casualties. First Eindecker victory was achieved on 1 August 1915 by Leutnant Max Immelmann. The 'scourge' ended only with the introduction into service of new Allied aircraft such as the RFC's DH2.

The first air raid on London was by Zeppelin LZ38 on 31 May 1915. The Kaiser had author-ised bombing of London, east of the Tower, a few days before and on the night of the 31st 3000 lb (1360 kg) of bombs were dropped on north-east London, killing seven people.

The first airship to be brought down by air attack was Zeppelin LZ37 on the night of 6/7 June 1915. In company with LZ38 and LZ39, the airship set out from Bruges to bomb London but adverse weather later forced them to alter course for their secondary targets – railways in the Calais area. LZ37 was located and attacked by Flight Sub-Lieutenant R A J Warneford of No 1 Squadron, RNAS, flying a Morane-Saulnier Parasol from Dunkirk. Warneford's only means of attack were six 20 lb (9 kg) bombs; he followed the airship from Ostend to Ghent, being forced to keep his distance by fire from the airship's gunners. He made a single pass over the airship dropping all six bombs from about 150 ft (45 m) above it. The sixth exploded, and the airship fell in flames on a suburb of Ghent killing two nuns. Only one member of Ober-leutnant Otto van de Haegen's crew survived. Warneford returned safely to base after making

Rare photograph of a Fokker Eindecker in flight during the 'Fokker Scourge' of 1915–16 (Imperial War Museum)

DH2, the RFC's first single-seat fighter

a forced landing to repair a broken fuel line. He was informed the following evening that he had been awarded the Victoria Cross; he died 12 days later when the tail of a Henry Farman pusher biplane collapsed in mid-air.

The first single-seat fighter to enter service with the RFC was the Airco (de Havilland) DH2, the prototype of which appeared in 1915. Powered by a 100 hp Gnome rotary engine, mounted in a 'pusher' configuration, the DH2 had a maximum speed of 93 mph (150 km/h) and was armed with one forward-firing Lewis machine-gun. It entered service with the RFC in early 1916, and was one of the fighters which ended the supremacy of the Fokker Eindecker. DH2s served, latterly in Palestine, until mid-1917. About 400 were built.

The first air attack using a torpedo dropped by an aeroplane was carried out by Flight Commander C H Edmonds, flying a Short 184 seaplane from HMS *Ben-My-Chree* on 12 August 1915, against a 5000 ton (5080 tonne) Turkish supply ship in the Sea of Marmara. Although the enemy ship was hit and sunk, the captain of a British submarine claimed to have fired a torpedo simultaneously and sunk the ship.

It was further stated that the British submarine *E14* has attacked and immobilised the ship four days earlier. However on 17 August 1915 another Turkish ship was sunk by a torpedo of whose origin there can be no doubt. On this occasion Flight Commander C H Edmonds, flying a Short 184, torpedoed a Turkish steamer a few miles north of the Dardanelles. His formation colleague, Flight-Lieutenant G B Dacre, was forced to land on the water owing to engine trouble but, seeing an enemy tug close by, taxied up to it and released his torpedo. The tug blew up and sank. Thereafter Dacre was able to take off and return to the *Ben-My-Chree*.

The first sustained strategic bombing offensive was opened by Italy on 20 August 1915, following its declaration of war against Austria–Hungary on 24 May. Major aircraft type used in the early raids was the Caproni Ca 2 three-engined biplane (100 hp Fiat A10s), of which 31 were delivered in 1915 and 133 in 1916. The Ca 2 was used in the first Italian night bombing raids. It carried a crew of four.

The first launching of an aeroplane by catapult on board ship (excluding anchored barge), took place on 5 November 1915 when an AB2 flying-boat was catapulted from the stern of the American battleship USS *North Carolina*, anchored in Pensacola Bay, Florida.

The first major fleet battle in which an aeroplane was used was the Battle of Jutland on 31 May 1916, when Flight-Lieutenant F J Rutland (accompanied by his observer, Assistant Paymaster G S Trewin) spotted and shadowed a force of German light cruisers and destroyers. Taking off from alongside HM seaplane carrier *Engadine* at about 15.10 h, Rutland

Short 184 seaplane

Curtiss flying-boat being catapulted from USS *North Carolina*, 5 November 1915

sighted the enemy ships and continued to radio position reports to the *Engadine*.

The first American pilot to be killed in the First World War was Victor Emmanuel Chapman of the Lafayette Escadrille, who was shot down near Verdun on 23 June 1916.

The first German airship to be brought down on British soil was the Schutte-Lanz SL XI, which was attacked on the night of 2 September 1916 by Lieutenant W Leefe Robinson, RFC, using the newly-invented Pomeroy incendiary ammunition. It crashed in flames near Cuffley, Hertfordshire. Robinson was awarded the VC.

The first radio-guided flying-bomb was tested on 12 September 1916. It was called the 'Hewitt-Sperry biplane' and was built by Curtiss. Powered by a 40 hp engine, it was capable of covering 50 miles (80 km) carrying a 308 lb (140 kg) bomb-load.

The first submarine to be sunk by an aeroplane was the French submarine *Foucault*, on 15 September 1916, by an Austrian Lohner flying-boat.

The first bombs to fall on London from an aeroplane were six small bombs dropped from a German LVG CII on 28 November 1916, falling near Victoria Station. The pilot of the attacking aircraft was Deck Offizier P Brandt.

The first British unit to be formed specifically for night bombing operations was No 100 Squadron, RFC, which formed at Hingham, Norfolk, in February 1917, and crossed to France on 21 March. A week later the unit received its first aircraft, 12 FE2bs, then being based at Saint-André-aux-Bois. Moving to Le Hameau on 1 April 1917, the squadron received four BE2es. The first operations were two raids on the night of 5/6 April 1917 on Douai Airfield, home base of the 'Richthofen Circus'. One FE2b failed to return; four hangars were badly damaged by bombs.

The first vessel in the world to be defined as an aircraft carrier (in the modern sense, i.e. equipped with a flying deck for operation of landplanes) was the light battle-cruiser HMS *Furious*. This ship commenced construction shortly after the outbreak of the First World War, it being intended to arm her with a pair of 18 in (457 mm) guns. In March 1917 authority to alter her design was issued, and at the expense of one of these huge guns she was completed with a hangar and flight deck on her forecastle. With a speed of 31·5 knots, she carried six Sopwith Pups in addition to four seaplanes. Her first Senior Flying Officer was Squadron Commander E H Dunning. HMS *Furious* became the **longest-lived active carrier in the world.** Between 1921 and 1925 the midships superstructure was eliminated and she emerged as a flush-deck carrier displacing 22 450 tons (22 809 tonnes), with two aircraft lifts and an aircraft capacity of 33. Her over-all length was 786 ft (239 m). After an extraordinarily active and exciting career in the Second World War (and a near head-on collision at night in the Atlantic with a troopship which passed so close as to carry away some of the carrier's radio masts), she was finally scrapped in 1949.

One of the longest-serving aircraft that was designed in the First World War was the two-seat Bristol Fighter. No fewer than 3101 were built during the war, the type first becoming operational on the Western Front in early April 1917. Further production was carried out until 1927. Some of the latter examples were equipped for use in tropical climates, and were used for patrol duty in countries like India, Iraq, Palestine and Egypt. Many were exported, serving in New Zealand up to 1936. The F2B Mk II version was powered by a 280 hp Rolls-Royce Falcon II engine, and had a maximum speed of 125 mph (201 km/h).

The worst month in terms of losses for the RFC was April 1917, 'Bloody April', when nearly 140 of the 365 RFC aircraft mustered for an offensive were lost in the first half of the month.

Gotha GIII bomber (Imperial War Museum)

The United States declared war on Germany on 6 April 1917, although the first American aerial patrol over enemy lines was not carried out until 19 March 1918.

The first German submarine to be sunk by an aeroplane was the German U-36, which was attacked in the North Sea by a Large America flying-boat piloted by Flight Sub-Lieutenant C R Morrish on 20 May 1917.

The first mass bombing raid on England by Gotha heavy bombers was made on 25 May 1917, when 21 aircraft attacked several towns including Folkestone and Shorncliffe. About 95 people were killed and many more injured.

The first German armoured aeroplane, designed for ground attack and low-level reconnaissance missions, was the Junkers J1, of which 227 were built. Fitted with 5 mm steel plating, the type entered service in mid-1917 and was armed with three machine-guns.

The most successful fighter aeroplane of the war was the Sopwith Camel, which achieved no fewer than 1294 victories over enemy aircraft. Production Camels were operated by the RFC and RNAS from mid-1917; a total of 5490 were built. Possessing excellent manoeuvrability, the Camel had a maximum speed of 115 mph (185 km/h).

The first mass bombing raid on London was by German Gotha heavy bombers on 13 June 1917. Fourteen bombers attacked an area around Liverpool Street Station, dropping 72 bombs and causing 162 deaths, with 432 persons injured. **This was the worst bombing raid of the war** in terms of dead and injured. **The last major bombing raid on England in daylight** was on 12 August 1917.

Squadron Commander E H Dunning's Pup going over the side of HMS *Furious*, 7 August 1917

Fokker D VII, Germany's best fighter of the war.

The first landing in the world by an aeroplane upon a ship under-way was carried out by Squadron Commander E. H. Dunning who flew a Sopwith Pup on to the deck of HMS *Furious* on 2 August 1917. Steaming at 26 knots into a wind of 21 knots, *Furious* thus provided a 47 knot headwind for Dunning who flew his Pup for'ard along the starboard side of the ship before side-slipping towards the deck located on the forecastle. Men then grabbed straps on the aircraft and brought it to a standstill. On 7 August Dunning attempted to repeat the operation in an even greater headwind but stalled as he attempted to overshoot and was killed when his aircraft was blown over the side of the ship.

The first Gotha bomber to be shot down at night during a bombing raid was destroyed in early 1918 by two Sopwith Camels of No 44 Squadron, RFC. This proved that even at night the Gotha could be intercepted, and night raids on England ceased in May 1918.

The first combat aeroplane to enter production in the United States was the British de Havilland DH4. The first machine was completed in February 1918, and by 5 November the same year 3431 had been completed. A total of 4846 was built before production stopped in 1919. The DH4 (or DH-4) was the only American-built aeroplane to fly over enemy territory during the First World War (which excludes of course the operations against Mexico in 1916).

The most successful German fighter of the war was the Fokker D VII, of which production versions were delivered from the Spring of 1918.

The type also served with several air forces after the war. Powered by a 185 hp BMW inline engine, it had a maximum speed of 124 mph (200 km/h) and excellent manoeuvrability. By the autumn of 1918, D VIIs equipped over 40 Jastas of the German air force.

The first flush-deck aircraft carrier in the world was HMS *Argus*, 15775 tons (16027 tonnes). Originally laid down in 1914 as the Italian liner *Conte Rosso*, she was purchased by Great Britain and launched in 1917, and completed in 1918. She featured an unrestricted flight deck of 565 ft (172 m) length and could accommodate 20 aircraft. She was ultimately scrapped in 1947. She was **the first carrier in the world to embark a full squadron of torpedo-carrier landplanes,** when in October 1918 a squadron of Sopwith Cuckoos was activated. They did not however see action.

HMS *Argus*

A squadron of DH4 day bombers of the RFC

Handley Page O/400 being towed at an operational aerodrome, with its wings folded

The first American-trained pilot to shoot down an enemy aircraft was Lieutenant Douglas Campbell, on 14 April 1918. He had received his training at an aviation school in America and was sent to the 94th Aero Squadron on 1 March 1918. In this squadron, with Captain Rickenbacker and Major Lufbery, he participated in the **first patrol over enemy territory by an American unit** on 19 March 1918. He was also **the first American, serving under American colours, to become an ace** by shooting down five enemy aircraft. His fifth German victim was shot down on 31 May 1918.

The Independent Force of the RAF was established on 5 June 1918 to carry out strategic bombing raids on German industrial and military areas. It was this force that dropped **the largest bombs of the war,** the 1650 lb (750 kg) 'block busters'. The first heavy bombers of the Force were Handley Page O/100s and O/400s, supplemented by the lighter Airco (de Havilland) DH4s, DH9s, DH9As and Royal Aircraft Factory FE2bs.

The first pilot to take off successfully from a towed barge in an aeroplane was the

Lieutenant Culley climbing into his lighter-borne Sopwith Camel, 11 August 1918 (Imperial War Museum)

American-born Flight Sub-Lieutenant Stuart Culley, RN, who on 1 August 1918 rose from a barge towed by HMS *Redoubt* at 35 knots. At 08.41 h on 11 August 1918 Culley took off from the barge while being towed off the Dutch coast and climbed to 18 000 ft (5500 m) to shoot down the German Zeppelin L53 using incendiary ammunition. He was thus **the first (and probably the only) pilot to shoot down an enemy aircraft after taking off from a towed vessel.** Landing in the sea alongside his towing destroyer, HMS *Redoubt*, he was rescued – and later awarded the DSO for his feat – and his Camel was salvaged by a derrick (invented by Colonel Samson). The only survivor of the Zeppelin baled out from 19 000 ft (5800 m) – almost certainly a record at that time.

The largest aeroplane force assembled during the war for a single military operation was that used during the battle for the Saint-Mihiel salient in the final weeks of the fighting. In command of about 1500 fighter, observation and bombing aircraft was General William Mitchell of the US Air Service.

The largest German aeroplane built during the First World War was the Aviatik R-Type (Riesenflugzeug) giant heavy bomber; basically a Zeppelin Staaken R VI built under licence by Automobil und Aviatik AG, this colossal aeroplane had a wing span of 180 ft 5½ in (55 m) and a length of 88 ft 7 in (27 m). It was powered by four 530 hp Benz Bz VI engines which gave it a maximum speed of 90 mph (145 km/h).

The last German airship attack on England which resulted in death or injury was made on 12 April 1918. Altogether, during the 51 Zeppelin airship raids on Great Britain during the war, 196 tons (199 tonnes) of bombs were dropped, killing 557 people and injuring many more.

During aeroplane bombing attacks on Great Britain during the war some 887 people were killed and over 2000 others injured.

THE GREAT AIR FIGHTERS OF THE FIRST WORLD WAR

The six most successful British and Empire pilots of the First World War

Major J T B McCudden, VC, DSO*, MC*,
MM, C DE G 57
Captain A W Beauchamp-Proctor, VC,
DSO, MC*, DFC 54
Captain D R MacLaren, DSO, MC*,
DFC, L D'H, C DE G . . . 54

*Bar to award

In addition to the above
8 pilots gained between 40 and 52 victories
11 pilots gained between 30 and 39 victories
57 pilots gained between 20 and 29 victories
226 pilots gained between 10 and 19 victories
476 pilots gained between 5 and 9 victories

Thus by the 'five victory' convention, the British and Empire air forces of the First World War produced 784 aces.

The six most successful German pilots of the First World War
Rittmeister Manfred, Freiherr von
Richthofen 80
Oberleutnant Ernst Udet . . . 62
Oberleutnant Erich Loewenhardt . . 53
Leutnant Werner Voss . . . 48
Leutnant Fritz Rumey . . . 45
Hauptmann Rudolph Berthold . . 44

All of these pilots were decorated with the Ordre Pour le Mérite.

In addition to the above
6 pilots gained between 40 and 43 victories
21 pilots gained between 30 and 39 victories
38 pilots gained between 20 and 29 victories
96 pilots gained between 10 and 19 victories
196 pilots gained between 5 and 9 victories

Thus by the 'five victory' convention, the Imperial German air forces of the First World War produced 363 aces.

The four most successful French pilots of the First World War
Capitaine René P Fonck . . . 75
Capitaine Georges M L J Guynemer . 54
Lieutenant Charles E J M Nungesser . 45
Capitaine Georges F Madon . . 41

In addition to the above
2 pilots gained between 30 and 39 victories
8 pilots gained between 20 and 29 victories
39 pilots gained between 10 and 19 victories
105 pilots gained between 5 and 9 victories
Thus the French air forces of the First World War produced 158 aces.

The four most successful American pilots of the First World War
Captain Edward V Rickenbacker, CMH,
DSC, L D'H, C DE G . . . 26
Second-Lieutenant Frank Luke Jr, CMH,
DSC, C DE G 21
Major G Raoul Lufbery, L D'H, MM,
C DE G, MC 17
Lieutenant G A Vaughn Jr, DSC, DFC . 13

In addition to the above
84 pilots gained between 5 and 12 victories; thus America produced during the First World War 88 aces. (It should be noted that the above figures include pilots who served with foreign air forces only, pilots who served with the American forces only, and pilots with mixed service, and all victories gained by these pilots irrespective of service.)

The four most successful Italian pilots of the First World War
Maggiore Francesco Baracca . . 34
Tenente Silvio Scaroni . . . 26
Tenente-Colonnello Pier Ruggiero Piccio 24
Tenente Flavio Torello Baracchini . 21

In addition to the above
39 pilots gained between 5 and 20 victories; thus Italy produced 43 aces during the First World War.

The four most successful Austro-Hungarian pilots of the First World War
Hauptmann Godwin Brumowski . 35–40
Offizierstellvertreter Julius Arigi . 26–32
Oberleutnant Frank Linke-Crawford 27–30
Oberleutnant Benno Fiala, Ritter von
Fernbrugg 27–29

(It should be noted that Austrian, Hungarian and Italian sources disagree as to the absolute accuracy of these pilot's scores.)

In addition to the above
Approximately 26 pilots gained between 5 and 19 victories. Thus it can be stated with reasonable certainty that the Austro–Hungarian Imperial air forces produced between 25 and 30 aces during the First World War.

The four most successful Imperial Russian pilots of the First World War
Staff Captain A A Kazakov, DSO, MC,
DFC, L D'H 17
Captain P V d'Argueeff . . . 15
Lieutenant Commander A P Seversky . 13
Lieutenant I W Smirnoff . . . 12

In addition to the above

Either 14 or 15 pilots gained between 5 and 11 victories; thus the Imperial Russian air forces produced either 18 or 19 known aces during the First World War. Other Russian pilots became aces, but the records are incomplete.

The four most successful Belgian pilots of the First World War

Second-Lieutenant Willy Coppens, DSO   37
Adjutant André de Meulemeester .   . 11
Second-Lieutenant Edmond Thieffry   . 10
Captain Fernand Jacquet, DFC   .   .   7

Confirmation of aerial victories during the First World War was subject to the most stringent regulations, and this has led to confusion over the actual number of victories achieved by various pilots. The figures quoted earlier are, with certain exceptions, those accepted officially as accurate in the countries of origin, and refer only to confirmed victories within the letter of the regulations. They are thus more liable to err on the side of under- rather than overstatement. Where certain notable pilots are considered to have destroyed more enemy aircraft than are allowed in their official totals, such unconfirmed figures are quoted below in the sections dealing with the pilots by name, or in the captions to the accompanying colour paintings.

The greatest ace of the First World War, in terms of confirmed aerial victories, was Rittmeister (Cavalry Captain) Manfred, Freiherr von Richthofen – the so-called 'Red Baron'. The eldest son of an aristocratic Silesian family,

Albatros D IIIs of the Richthofen 'Circus'

he was born on 2 May 1892 and was killed in action on 21 April 1918, by which time he had been credited with 80 victories, had been awarded his country's highest decoration, commanded the élite unit of the Imperial German Air Service (Luftstreitkräfte), and was the object of universal adulation in his homeland and an ungrudging respect among his enemies. Early in the war Richthofen served on the Eastern Front as an officer in Uhlan Regiment Nr 1 'Kaiser Alexander III', and transferred to the Air Service in May 1915. His first operational posting was to Feldfliegerabteilung Nr 69; with this unit he flew two-seater reconnaissance machines in the East – without apparently any unusual skill. In September 1916 he was selected for Jagdstaffel 2, the scout squadron led by Oswald Boelcke (q.v.). His first officially recognised victory was over an FE2b of No 11 Squadron, RFC; Richthofen, flying an Albatros D II scout, shot down this aircraft on 17 September 1916; the crew, Second-Lieutenant L B F Morris and Lieutenant T Rees, lost their lives. Richthofen continued to score steadily, and in January 1917 was awarded the coveted 'Blue Max', the Ordre Pour le Mérite. He was given command of Jagdstaffel 11, and characteristically maintained a collection of silver cups, each engraved with the particulars of a victim. The silversmith's most lucrative month was the 'Bloody April' of

Rittmeister Manfred, Freiherr von Richthofen, head bandaged after a recent wound, with Kaiser Wilhelm II

1917, when Richthofen shot down 21 aircraft. In June 1917 he was given command of a new formation, Jagdgeschwader Nr 1, comprising Jastas 4, 6, 10 and 11; this group of squadrons became known to the Allies as 'Richthofen's Flying Circus', because of the bright colours of their aircraft. Contrary to popular legend Richthofen did not invariably fly a personal all-red aircraft but a variety of Albatros D IIIs and Fokker Dr Is, some of which were painted blood-red all over and some only partially red. Richthofen's death on 21 April 1918 has been the subject of controversy ever since. He was flying Fokker Dr I number *425/17* when he became engaged in combat with Sopwith Camels of No 209 Squadron, RAF, over Sailly-le-Sec. At one point, Second-Lieutenant W R May was flying at low altitude with Richthofen in pursuit and the aircraft of Captain A Roy Brown, DSC, diving to attack the German. Brown opened fire in an attempt to save the inexperienced May from the enemy ace, and Richthofen's triplane was then seen to break away and crash-land. Richthofen was found dead in his cockpit with a bullet wound in the chest. At about the same time as Brown attacked, machine-gunners of an Australian Field Artillery battery fired at Richthofen's aircraft. Although Brown was officially credited with the 'kill', it has never been established who fired the fatal shot.

The first true fighter leader of the First World War was Hauptmann Oswald Boelcke, whose name, with those of Richthofen and Immelmann, is still commemorated today in the honour title of a German Air Force combat unit. Boelcke was born on 9 May 1891 and was commissioned in a communications unit in 1912. He became interested in aviation during army manoeuvres, and gained his Pilot's Certificate at the Halberstadt Flying School on 15 August 1914. He was posted to La Ferte to join Feldfliegerabteilung No 13 in September, and, with his brother Wilhelm as observer, soon amassed a considerable number of sorties in Army Co-operation Albatros B II biplanes. By early 1915 he had 42 missions in his log-book, and had been awarded the Iron Cross, Second Class. The visit of Leutnant Parschau to his unit to demonstrate the Fokker M8 monoplane scout fired him with enthusiasm; and in April, having received the Iron Cross, First Class, he secured a posting to Hauptmann Kastner's Feldfliegerabteilung No 62, where he flew an armed machine for the first time – an Albatros C I, number *162/15*. Before long, he was selected to fly early examples of Fokker's E-series armed monoplane scouts; few were available, and Boelcke, Kastner and Leutnant Max Immelmann at first took turns to fly them. His success as a combat pilot was matched by his grasp of technical matters and his organising ability. In particular, his ideas for the use of squadrons composed entirely of single-seat fighting scouts commanded attention in high places; until mid-1916 most units operated mixed equipment. After a tour of other fronts in early 1916, Boelcke returned to the West and was given command of the new Jagdstaffel Nr 2 (Jasta 2), which was equipped with Albatros D I and D II scouts. Boelcke was killed on 28 October 1916 during an engagement in which one of his colleagues, Leutnant Boehme, who was flying close to him, banked sharply. Boehme's undercarriage struck the wing of his Albatros, which spiralled to the ground. He was 25 years old, a holder of the Ordre Pour le Mérite and numerous other decorations, the victor of 40 aerial combats, and the idol of his country.

One of Germany's first two great fighter aces was Leutnant Max Immelmann, 'The Eagle

Max Immelmann

of Lille'. He was serving with Feldfliegerabteilung No 62 at Douai when the first Fokker monoplane scouts became available. Hauptmann Kastner instructed Boelcke in the subtleties of the new machine, and Boelcke taught Immelmann. On 1 August 1915 Immelmann scored his first victory while flying an E I (believed to have been one of Boelcke's two machines) when his comrade was forced to drop out of the fight with a defective machine-gun. Thereafter he and Boelcke ranged over their sector of the front, sometimes together, sometimes alone, hunting the enemy from the sky. Even a year later only a comparative handful of German pilots were operating the Fokker scouts; yet so great was the superiority of the agile single-seater, with its synchronised forward-firing machine-gun, that the 'Fokker Scourge' became a major disaster for Allied arms. Immelmann met his death, after shooting down 15 Allied aircraft, on 18 June 1916. Flying near Lens, he attacked an FE2b of No 25 Squadron, RFC. Another FE, flown by Second-Lieutenant G R McCubbin, with Corporal J H Waller as gunner, attacked him. The Fokker made an attacking pass, then went into a dive and broke up in mid-air. Some sources claim that his death was caused by technical failure, but the RFC credited Corporal Waller with the victory.

The first great British ace, Captain Albert Ball occupied a place in the affections of the British public and armed forces analogous to that held by Max Immelmann in Germany. He was the first high-scoring fighter pilot whose exploits became widely known on the home front, his success stemming partly from his practice of charging at the enemy whether equally matched or heavily outnumbered. Born in Nottingham on 14 August 1896, Albert Ball joined the Sherwood Foresters on the outbreak of war. During a visit to Hendon he became interested in flying, and secured a transfer to the RFC. He joined No 13 Squadron in France on 15 February 1916, and flew BE2cs on artillery-spotting flights. In May he was posted to No 11 Squadron, which had in charge a Nieuport scout. He immediately became attached to this aircraft, and was to fly Nieuports by preference throughout most of his career. His first two successes came on 22 May, when he drove down (but could not get confirmed) an Albatros D I, and forced an LVG two-seater to land. On 27 June his MC was gazetted; and on 2 July he shot down his first Roland C II. Given a new Nieuport on his return to No 11 from 'rest' with No 8 Squadron, on 10 August, Ball resumed his private war against the Roland C IIs, which were his favourite prey. When homogeneous fighting squadrons were formed he took his Nieuport to No 60 and was given a roving commission, which suited his style admirably. Uncaring of odds, he would charge at enemy formations and deliver a devastating fire at close range, generally from a position immediately below the belly of the enemy machine, with his wing-mounted Lewis gun pulled down and back to fire almost vertically upwards. His DSO and Bar were gazetted simultaneously on 26 September, and a second Bar on 25 November, after he destroyed or forced down five Rolands and five Albatros's between 15 and 28 September 1916. By the time he left France on 4 October Ball was credited with the destruction of ten enemy aircraft and with forcing down 20 more. On 7 April 1917, after a period spent instructing pupil pilots in England, he returned to the front as a Flight Commander in No 56 Squadron. This unit flew the new SE5 scout, although Ball also flew a Nieuport by choice. His 47th and last victory, on 6 May 1917, was against an Albatros scout of Jasta 20. Later the following evening Ball, flying his SE5, dived into dense cloud while chasing a German single-seater near Lens; the enemy later discovered his wrecked aircraft and his body. His death remains a mystery. Lothar von Richthofen was officially credited with the victory, but himself denied it, maintaining that the aircraft he shot down was a triplane – an opinion confirmed by other witnesses. Ball's body bore no wound, and what caused his aircraft to crash has never been established. He was 20 years and nine months old when he died; his Victoria Cross was gazetted on 3 June 1917.

The greatest Allied ace of the First World War was Capitaine René Paul Fonck, who served with Escadrille SPA103, one of the units of the famous Groupe de Combat No 12 'Les Cigognes'. Officially Fonck is credited with 75 victories; his own personal estimate, including aircraft destroyed but not confirmed by Allied ground observers, was 127. Born in the Vosges in 1894, Fonck died peacefully in his sleep at his Paris home on 18 June 1953. A keen aviation enthusiast in his boyhood, Fonck was disgusted to find himself posted to the 11th Engineer Regiment on mobilisation in 1914. After un-

happy months digging trenches, he finally reported to Saint-Cyr for aviation training in February 1915; and in June, having received his Brevet, he joined Escadrille C47. He flew Caudron G IVs on reconnaissance duties and low-level bombing missions, and distinguished himself by his courage. In July 1916 he fitted a machine-gun to his aircraft and on 17 March 1917 he fought off five Albatros's, destroying one. This second confirmed victory (the first being a forced-down Rumpler on 6 August) led to his transfer to the 'Cigognes' group a month later. On 9 May 1918 he achieved no fewer than six confirmed 'kills' – including three two-seaters destroyed in 45 s, the three wrecks being found in a radius of 1200 ft (365 m). On 26 September he again shot down six aircraft, comprising a two-seater, four Fokker D VIIs, and an Albatros D V. Fonck's last victory was over a leaflet-dropping two-seater on 1 November 1918. He was a thoughtful and analytical pilot, a master of deflection shooting, and his economy of ammunition bordered on the uncanny; he frequently sent an aircraft down for the expenditure of only five or six rounds, placed, in his own words, 'comme avec la main' – 'as if by hand'. His many decorations included the Belgian Croix de Guerre, the French Croix de Guerre with 28 palms, the British Military Cross and Bar, the British Military Medal and the Cross of Karageorgevitch; his first sextuple victory also brought him the Croix d'Officier de la Légion d'Honneur. There is little doubt that he was **the most successful fighter pilot of any combatant nation in the First World War.**

Britain's most successful fighter pilot in the First World War was Major Edward 'Mick' Mannock. His score of combat victories stands at 73, but he is known to have insisted that several additional victories justly attributable to him should be credited to other pilots. Born on 24 May 1887, the son of a soldier, Mannock was working in Constantinople when the war broke out, and was interned by the Turks. He was repatriated in April 1915 on health grounds and rejoined the Territorial Army medical unit to which he had belonged before leaving the country. He was commissioned in the Royal Engineers on 1 April 1916 and finally transferred to the Royal Flying Corps in August 1916. His acceptance for flying duties was remarkable as he suffered from astigmatism in the left eye, and must have passed his medical by a ruse. He gained his Pilot's Certificate on 28 November

Major Edward 'Mick' Mannock (Imperial War Museum)

1916, and was posted to No 40 Squadron, France, on 6 April 1917, the unit being equipped at that time with Nieuport scouts. He shot down a balloon on 7 May, and on 7 June scored his first victory over an aeroplane. Returning from leave in July he shot down two-seaters on the 12th and 13th of that month, and his Military Cross was gazetted. He was promoted Captain, and took command of a flight. His score grew rapidly, as he was possessed by a bitter and ruthless hatred of the enemy uncommon among his contemporaries. His care of the pilots under his command, however, was irreproachable, and he has been judged the greatest patrol leader of any combatant air force. In January 1918 he returned to England to take enforced leave, by which time his score stood at 23. He returned to France in March as a Flight Commander in the newly formed No 74 ('Tiger') Squadron, equipped with the SE5a, and in his three months with the unit added 39 to his score. He was promoted Major in mid June, and was given leave before taking command of No 85 Squadron. With No 85 he raised his score to 73 before being shot down by German ground fire that hit his petrol tank on 26 July. His grave has never been found and it was nearly a year later that he was awarded a posthumous Victoria Cross.

France's second most successful fighter pilot was Capitaine Georges Marie Ludovic

Jules Guynemer, who served in the 'Cigognes' group with Escadrille N3/SPA3 and achieved 54 confirmed victories. Guynemer was born on Christmas Eve 1894. A frail and delicate youth, he was twice rejected for military service before finally securing a posting to Pau Airfield as a pupil-mechanic in November 1914. A transfer to flying training followed, and on 8 June 1915 he joined Escadrille MS3, then flying Morane Bullet monoplanes. (In the French manner, the unit was later designated N3 and SPA3 on re-equipment with Nieuport and SPAD scouts respectively.) Guynemer's first victory came on 19 July 1915; by July the following year he had eight victories to his name, and was flying Nieuports. During the latter half of 1916 his score quickly increased, bringing decorations and promotion, and by the end of January 1917 he was credited with 30 kills. A quadruple victory (two within the space of one minute) on 25 May brought his score to 45. He continued to fly combat sorties despite failing health and attempts to ground him, and was shot down seven times. He failed to return from a flight over Poelcapelle (Belgium) on 11 September 1917, and was mourned by his whole nation. No trace of his aircraft or his body has ever been found. No German pilot put in an immediate claim, but a certain Leutnant Wisseman was rather belatedly credited with his death by the German authorities. The claim is open to question, and Wisseman was killed by René Fonck three weeks later.

The third of France's First World War aces was Lieutenant Charles Eugène Jules Marie

Capitaine Georges Guynemer in his Nieuport 23 with 'Stork' insignia, 1916

with N65 opened with a victory on 28 November, and continued in a series of brilliant successes punctuated by frequent spells in hospital. Nungesser suffered numerous wounds and injuries, and was particularly dogged by the aftermath of a serious crash in January 1916; he sustained multiple fractures which failed to knit satisfactorily, and for the rest of the war was obliged to make periodic trips to hospital to have the bones re-broken and re-set. Despite the appalling pain from which he was seldom free Nungesser continued to fly and to add to his score. Many of his total of 45 kills were gained during a period when he was unable to walk, and had to be carried to and from his aircraft for every flight. After the war he turned first to running a flying school, and later to barnstorming in the United States. Finally he became absorbed in the idea of an East–West Atlantic attempt, and the Levasseur Company prepared a special aircraft designated the 'PL8'. It was named *Le Oiseau Blanc* – 'The White Bird' – and bore on the fuselage the macabre insignia which Nungesser had made famous in the skies over the Western Front – a black heart charged with a skull and crossbones, two candles and a coffin. With Capitaine Coli as navigator, Nungesser flew 'The White Bird' out over Le Havre at 06.48 h on 8 May 1927, and was never seen again.

The second most successful British and Empire pilot of the war was a Canadian, William Avery Bishop, born on 8 February 1894 in Ontario. While in England as a cavalry subaltern in the Canadian Mounted Rifles in 1915, Bishop decided he would see more action as a pilot, and transferred to the Royal Flying Corps in July of that year. He flew in France as an observer with No 21 Squadron for several months, and was hospitalised as the result of a crash-landing and frostbite. He trained subsequently as a pilot and joined No 60 Squadron in March 1917. The squadron was at that time equipped with Nieuport 17 scouts, an aircraft which Bishop was to handle brilliantly. On 25 March he scored his first victory over an Albatros and subsequently gained many honours including the Victoria Cross for his action over an enemy airfield on 2 June. When his score reached 45 Bishop was promoted Major and awarded a Bar to his DSO. Late in 1917 and early in 1918 he carried out a number of non-combat duties, including recruiting drives in Canada and instructing at an aerial gunnery school. He was subsequently given command of

Lieutenant Charles Nungesser (Imperial War Museum)

Nungesser of Escadrilles VB106 and N65. A brilliant athlete and promising scholar, he had left France as a teenage boy and sailed to South America to find an uncle who was reputed to live in Rio de Janeiro. Failing to locate him, Nungesser got a job and stayed in South America for several years, winning a name as a racing motorist and teaching himself to fly. He returned to France in 1914, joined the 2nd Hussar regiment, and distinguished himself in a lone battle against a squad of enemy infantry and a car full of German officers during the Battle of the Marne. He then transferred to the Service Aéronautique, and reported to the Voisin reconnaissance and bombing unit VB106 on 8 April 1915. On 26 April he was shot down, and for the next two months devoted himself to vain attempts to lure enemy scouts within range of his lumbering Voisin – usually by imitating a crippled aircraft – in the hope of exacting revenge. He finally scored his first victory late in 1915, on an unauthorised night sortie, gaining himself the Croix de Guerre and eight days' detention. In November he joined Escadrille N65. His career

Raymond Collishaw, brilliant exponent of the
Sopwith Triplane and leader of the formidable
'Black Flight', who achieved sixty confirmed
victories (Imperial War Museum)

No 85 Squadron, flying SE5as, and went back
to France on 22 May 1918. After gaining 27 more
victories, he was recalled to England, and never
flew operationally again. His DFC was gazetted
on 2 July. Bishop remained in the service, rising
to the rank of Honorary Air Marshal in the
Royal Canadian Air Force. He died in Florida,
USA, in September 1956.

**The most successful fighter pilot of the
Royal Naval Air Service** during the First
World War, and, with 60 confirmed victories,
third in the over-all British and Empire aces' list,
was Raymond Collishaw. Born on 22 Novem-
ber 1893 at Nanaimo, British Columbia, he
went to sea at the age of 17 and served as a
Second Mate on a merchant ship, and later on
fishery protection vessels. He transferred to the
RNAS in January 1916, joining No 3 Wing in
August, and scored his first victory on 12 October. In February 1917 he joined a scout unit, No
3 (Naval) Squadron, and in April was posted to
No 10 (Naval) Squadron as commander of 'B'
Flight. Equipped with Sopwith Triplanes, the
'Black Flight' of 'Naval Ten' earned a reputation
as one of the most formidable Allied units of the
war. The Flight was composed entirely of Canadians; their aircraft were decorated with black

paint, and named *Black Maria* (Collishaw), *Black
Prince, Black Sheep, Black Roger* and *Black Death*.
Between May and July 1917 the Flight destroyed
87 enemy aircraft, and during June Collishaw
himself shot down 16 in 27 days. On 6 June he
shot down three Albatros's in a single engagement, for which feat he was awarded the DSC.
By 3 July, his score stood at 28, and Collishaw
gained the DSO. He shot down ten more victims before the end of the month and was shot
down himself, for the second time without
significant injury. After home leave Collishaw
returned to France on 24 November 1917, taking command of No 13 (Naval) Squadron, a
Sopwith Camel unit. He was posted back to No
3 (Naval) Squadron as commander in January
1918, with a total of 41 victories. He was back
in England on administrative duties in October,
after reaching a total of 60 victories; but commanded No 47 Squadron in the Russian campaign of 1919/20, where he destroyed two more
aircraft. He remained in the Royal Air Force,
serving in the Second World War and reaching
the rank of Air Vice-Marshal, CB, with the
DSO and Bar, DSC, DFC and Croix de Guerre,
as well as both military and civil grades of the
OBE.

**The most successful American pilot of the
First World War** was Captain Edward Vernon
Rickenbacker, with 26 confirmed aerial victories. Born on 8 October 1890 in Columbus,
Ohio, Rickenbacker made a considerable name
for himself between 1910 and 1917 as one of
America's leading racing motorists. While in
England in 1917, he became interested in flying
and when America entered the war he returned
home and advanced the idea of a squadron composed entirely of racing drivers. The idea did not
arouse official interest, but a meeting with General Pershing in Washington led to Rickenbacker's enlistment and sent him to France as
the General's chauffeur. In August 1917 he transferred to the Aviation Section, and his mechanical expertise led to a posting to the 3rd Aviation
Instruction Center at Issoudun as Chief Engineering Officer. In his own time he completed
advanced flying and gunnery courses, and on
4 March 1918 finally secured a transfer to the
94th Aero Squadron – the 'Hat-in-the-Ring'
squadron commanded by Raoul Lufbery, the
Escadrille Lafayette ace. With Lufbery and
Douglas Campbell, Rickenbacker flew the first
American patrol over enemy lines on 19 March,
and on 29 April he shot down his first victim,

Above: Blériot XI monoplane of 1913, in the Swiss Transport Museum, Lucerne.

Right: Faithful replicas of many great combat aircraft of the First World War can be seen in the air in every part of the world. This Sopwith Pup was built in the USA (*James Gilbert*).

Below: Boeing's P-26 'Peashooter', first monoplane fighter built for the US Army Air Corps, was still in first-line service in the Philippines when America entered the Second World War. All those based near Luzon were promptly lost to enemy action or accidents. This one served in the Panama Canal Zone, was presented to Guatemala in 1942, returned to America in 1958, and is now displayed at the USAF Museum in the colours of the 34th Attack Squadron.

Above: About 1500 Lockheed Hudsons were bought for the Royal Air Force at a time when the British Purchasing Commission was keeping the US aircraft industry far more busy than was the US Army Air Corps. This one was retained in the USA as an AT-18 gunnery trainer.

Left: With a large US flag painted on its side, to proclaim its neutrality, Pan American Boeing 314 *American Clipper* is prepared for a transatlantic flight early in the Second World War.

Below: Boeing Model 247D, still flying as a reminder of the type which was the ancestor of all modern monoplane airliners with a retractable undercarriage.

Refuelling a Boeing B-17F Flying Fortress, spearhead of the US daylight bomber offensive against Europe.

Designed as America's counterpart of the B-17 for the remaining years of the 20th century, the Rockwell International B-1 here demonstrates its ability to make use of natural cover during low-altitude penetration at near the speed of sound.

First airliner to utilise Rolls-Royce's highly advanced RB.211 turbofan engine, a Lockheed TriStar wide-bodied transport.

Inca symbols, strange beasts and huge red sunbursts on the engine pods identify the Braniff DC-8-62 *Flying Colours of South America* decorated in 1973 by artist Alexander Calder.

Douglas Skyhawks of the US Navy's famous *Blue Angels* aerobatic team.

Top: Lockheed C-130E Hercules of the Egyptian
Air Force, over the Pyramids. More than 1500
Hercules have been built for military use.

Centre: British-built Lightning F.53 interceptor/
ground-attack fighters of the Royal Saudi Air Force.

Left: Mirage 5 of the Pakistan Air Force, built in
France by Dassault.

Right: Prototype of the Italian Aermacchi MB339 trainer/ground-attack aircraft, successor to the MB326 used by many world air forces.

Centre: A Harrier V/STOL attack/reconnaissance aircraft of No 20 Squadron, Royal Air Force, comes in to land during Exercise Grimm Charade in Germany, while a Harrier of 3 Squadron is towed to a camouflaged dispersal area (*Ministry of Defence*).

Below: Hawker Hunter of No 45 Squadron, Royal Air Force.

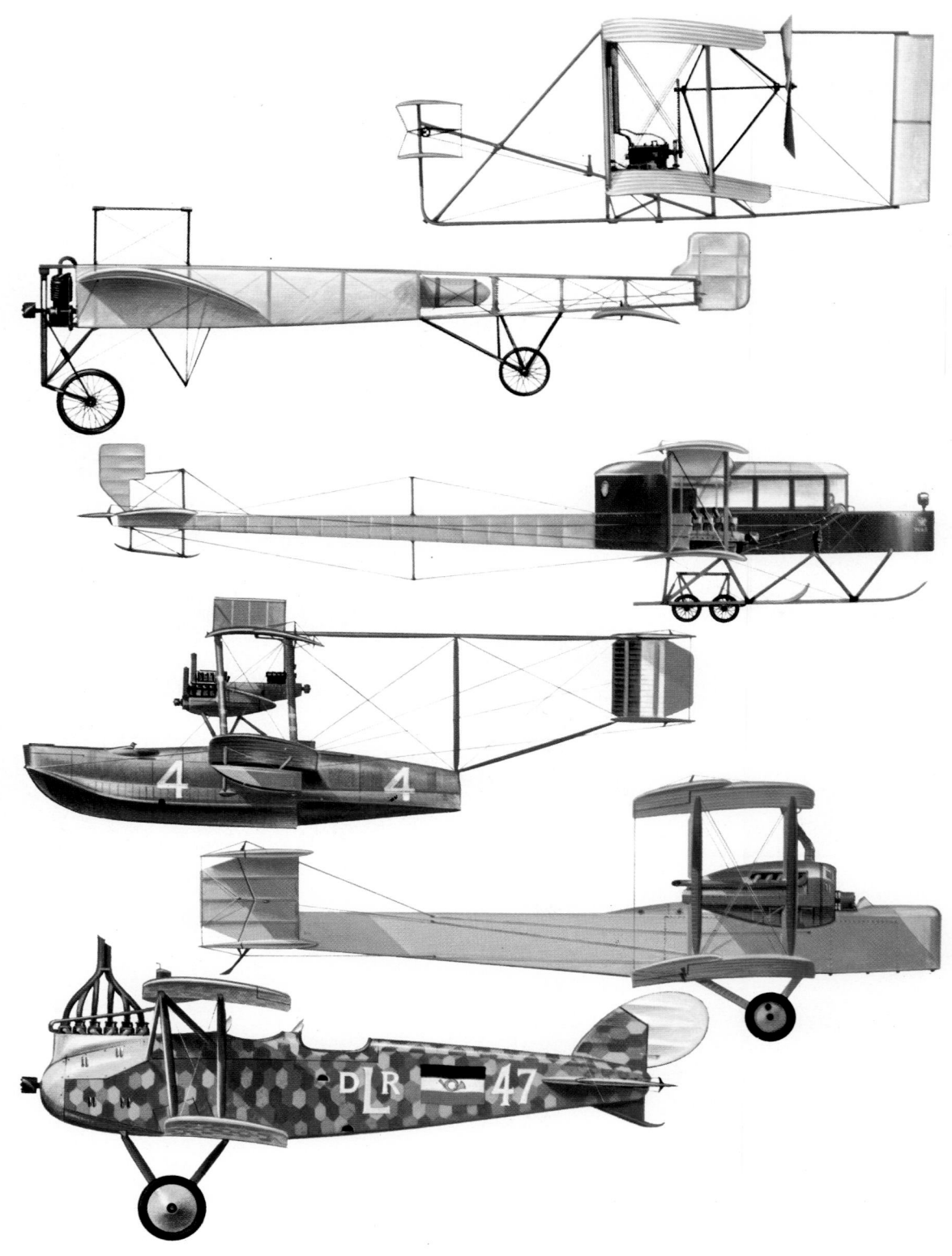

The Wright Flyer *of 1903 is recognised throughout the world as the first aeroplane to have made a powered, controlled and sustained flight. Built by Orville and Wilbur Wright, cycle-making brothers of Dayton, Ohio, USA, it flew for the first time at 10.35 am on 17 December 1903, at Kill Devil Hills, Kitty Hawk, North Carolina. Length of that first flight was 120 ft (36·5 m), which is less than the wing span of most modern airliners. The first Flyer logged three further flights on that same day, then was blown over by the wind and damaged, and never flew again. Its total flying time was 98 seconds.*

This Blériot XI *monoplane made the first crossing of the English Channel by an aeroplane, by flying from Les Baraques, near Calais, to Northfall Meadow, by Dover Castle, in 36½ minutes on 25 July 1909. More than any previous aviation exploit, this flight made the public aware of the future possibilities of international air travel. Designed and flown by Louis Blériot, the Type XI monoplane had a wing span of 25 ft 7 in (7·80 m), length of 26 ft 3 in (8·00 m), loaded weight of 661 lb (300 kg) and speed of 36 mph (58 km/hr). It was powered by a 25 hp three-cylinder Anzani engine.*

Igor Sikorsky's huge biplane, built in Russia in 1912–13, was officially named Russian Knight *but is usually remembered as the* Bolshoi *(Grand) because of its great size. First four-engined aeroplane to fly, on 13 May 1913, and the first to have a luxuriously furnished passenger cabin and washroom, it was powered by 100 hp Argus engines which gave it a cruising speed of about 55 mph (88 km/hr). Wing span was 91 ft 11 in (28·00 m), length 65 ft 8 in (20·00 m) and loaded weight about 9000 lb (4080 kg). The* Bolshoi *flew 53 times before being damaged by an engine which fell off a crashing aircraft, after which it was dismantled. It could carry eight persons.*

The Curtiss flying boat NC-4 *was the first aircraft to cross the Atlantic. It was one of four similar large machines ordered by the US Navy during the First World War, and was launched at Far Rockaway, New York, on 30 April 1919. NC-2 was dismantled to provide spares for NC-1, NC-3 and NC-4, which took off from Trepassy, Newfoundland, on 16 May 1919. NC-1 and NC-3 were both forced down at sea. NC-4, commanded by Lt Cdr A C Read, reached Horta in the Azores on 17 May, Ponta Delgada on the 20th, Lisbon on the 27th and Plymouth on the 31st. Powered by four 400 hp Liberty engines, NC-4 spanned 126 ft (38·40 m), had a length of 68 ft 3½ in (20·85 m), loaded weight of 28 500 lb (12 925 kg) and maximum speed of 91 mph (146 km/hr).*

This Vickers Vimy, *still to be seen in the Science Museum, London, made the first-ever non-stop transatlantic flight, flown by Capt John Alcock and Lt Arthur Whitten Brown. Both men were knighted for the achievement, which also won a Daily Mail prize of £10 000. Powered by two 360 hp Rolls-Royce Eagle VIII engines and modified to carry 865 gallons (3392 litres) of fuel, the Vimy covered the 1890 miles (3040 km) from Newfoundland to Ireland in just under 16 hours on 14–15th June 1919. It had a span of 67 ft (20·42 m), length of 42 ft 8 in (13·00 m), weight of 13 000 lb (6033 kg) at take-off in Newfoundland, and normal maximum speed of 100 mph (161 km/hr).*

Wartime LVG C VI biplanes *were used on the world's first sustained daily passenger services, opened between Berlin and Weimar in Germany, via Leipzig, by Deutsche Luft-Reederei on 5 February 1919. The C VI was normally powered by a 200 hp Benz Bz IV engine, had a loaded weight of about 3086 lb (1400 kg), accommodation for a pilot and two passengers in open cockpits, a speed of just under 100 mph (161 km/hr) and span of 42 ft 7¾ in (13·00 m).*

Fokker D VII *flown during the spring of 1918 by Hauptmann Rudolf Berthold as commanding officer of Jagdgeschwader Nr 2. Sixth in the roll of German aces, with forty-four confirmed victories, Berthold suffered continual pain from an injury which rendered his right arm useless. He had the controls of his Fokker altered to compensate for his disability and continued to fly, scoring at least sixteen of his victories during a period when his arm refused to heal and was rejecting splinters of suppurating bone almost daily. Active in the Freikorps movement in the immediate post-war period, Berthold led an anti-Communist band known as the 'Eiserne Schar Berthold'. He was murdered – by strangulation with the ribbon of his Ordre Pour le Mérite – on 15 March 1920, after accepting a safe-conduct offer from a Communist group in Harburg.*

Siemens-Schuckert DIII *flown late in the summer of 1918 by Oberleutnant Ernst Udet as commanding officer of Jasta 4, based at Metz. This gifted pilot was credited with sixty-two victories, and in this respect was second only to Manfred von Richthofen; his amiable and rather flamboyant disposition made him a popular commander, although an early over-confidence sometimes led him into difficulties. He survived the war, and became well known as a stunt pilot, explorer and international aviation 'playboy'. His extrovert nature and long friendship with many leading personalities in German aviation led to senior appointments under the Nazi régime; but he eventually found himself playing a role for which he was temperamentally unsuited, and a growing sense of estrangement culminated in his suicide on 17 November 1941. The monogram displayed on his aircraft referred to his fiancée, Fräulein Lola Zink.*

Fokker Dr I, *number 152/17, sometimes flown by Rittmeister Manfred, Freiherr von Richthofen as commanding officer of Jagdgeschwader Nr 1 during the early months of 1918. (Details of Richthofen's career may be found in the body of the text.) Richthofen is known to have been flying this aircraft on 12 March 1918 when he gained the sixty-fourth of his eighty confirmed victories; his victim on that occasion was a Bristol F2B, number B1251, of No 62 Squadron, RFC brought down near Nauroy. The crew, Lieutenant L C F Clutterbuck and Second-Lieutenant H J Sparks, survived the crash and were made prisoners.*

Albatros D III *flown during the spring of 1917 by Leutnant Werner Voss, whose forty-eight confirmed victories place him fourth in the roll of German aces. Voss became a pilot in May 1916 and served his apprenticeship in the famous Jasta Boelcke; he was a born flyer and fighter, but lacked the leadership qualities of Richthofen and Berthold. During the spring of 1917 he is believed to have served briefly with both Jasta 5 and Jasta 14, and it is with the former unit that this Albatros is usually associated. In July 1917, at the request of his friend Richthofen, Voss was posted to command Jasta 10 in Jagdgeschwader Nr 1. During their lifetimes, Voss was second-ranking ace to Richthofen; he was finally killed on 23 September 1917 in a prolonged dogfight between himself (flying Fokker Dr I 103/17) and one Albatros scout against seven SE5as, including the whole of 'B' Flight, No 56 Squadron, RFC, led by Captain J T B McCudden. Voss was twenty years and five months old when he died.*

Fokker E I, *number 3/15, flown by Leutnant Oswald Boelcke of Fliegerabteilung 62, based at Douai in August 1915. (Details of Boelcke's career may be found in the body of the text.) It is believed that this was the machine flown by Leutnant Max Immelmann of the same unit when he gained his first victory on 1 August 1915.*

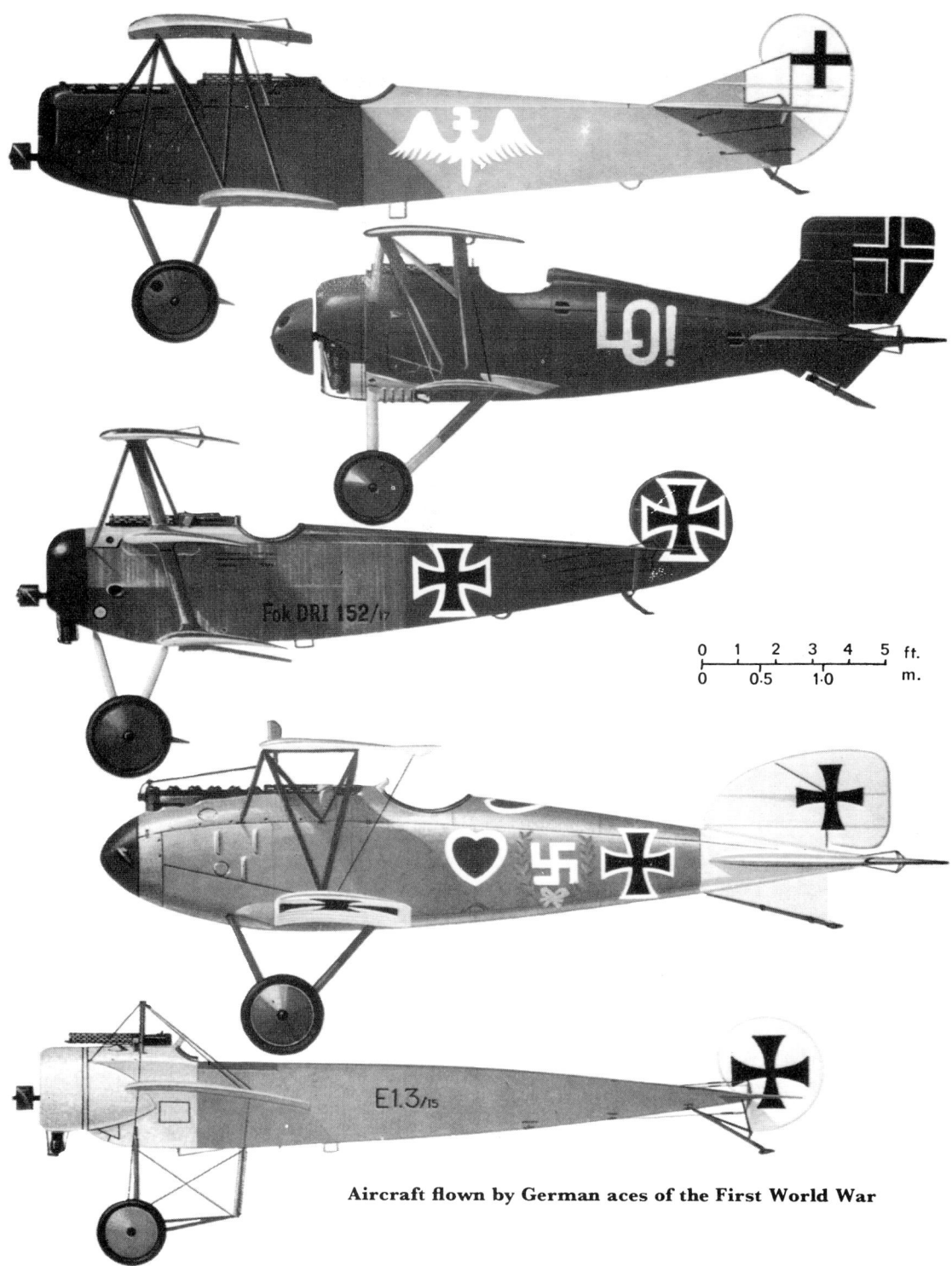

Fok DRI 152/17

E1.3/15

Aircraft flown by German aces of the First World War

an Albatros scout. On 30 May his fifth victory qualified him as an ace, but it was to be his last for four months. An ear infection put him in hospital and convalescence until mid September, when he returned to the squadron as a Captain and Flight Commander. He took over command of the 94th on 25 September, and continued to score heavily until the Armistice. Captain Rickenbacker was active in the automobile and airline industries between the wars, and was largely responsible for building up Eastern Air Lines, of which corporation he became Chair-

man in 1953. During the Second World War he toured widely, visiting Air Force units abroad and undertaking various missions for his Government. In the course of a flight over the Pacific his aircraft was forced to ditch, and Rickenbacker and the crew survived 21 days on a life-raft before being picked up. He remained active in various public fields until his death on 23 July 1973, at the age of 82. His many American and foreign decorations included his country's highest award for gallantry, the Congressional Medal of Honor.

Captain 'Eddie' Rickenbacker with his SPAD XIII

Mitsubishi 1MF1

SECTION 5

BETWEEN THE WARS - MILITARY

The ending of the First World War left Britain with the largest air service in the world, with more than 22 500 aircraft on strength. Within months this force was reduced to just a few squadrons, and surplus military aircraft were given away to other countries, scrapped or sold at minimal cost to private pilots or flying clubs. However, the most important change in the history of transportation was about to unfold, with the gradual acceptance of air travel by the public.

Virtually all civil flying in Europe had stopped during the war. Now, with the return of peace, flying became the new 'thing to do'. The abundance of pilots, and the availability of ex-military aircraft to convert into makeshift airliners, gave rise to small airline companies which, for a relatively handsome fee, were willing to transport anyone brave enough to anywhere within range. Some European operators began scheduled services to carefully chosen destinations at home and abroad; and as the first post-war decade progressed such services began to appear in every continent.

The world had become aviation-minded; but while civil flying expanded, due mainly to government support through generous air mail contracts, and to the availability of purpose-built airliners with improved accommodation and range, military aviation was having a rough time. Like many air forces, the RAF had to make do for several years with left-over wartime aircraft. It demonstrated for the first time the effectiveness of what we now call a 'deterrent' policy, based almost entirely on air power, when it was allocated the task of policing the Middle East and North-West Frontier of India. Eventually, to meet the demanding needs of increasingly adventurous airlines, and air forces operating all over the world, this period of 1918–1939 brought the general use of metal structures and skins, monoplane wings, enclosed cockpits and retractable undercarriages.

By the mid-1930s Germany had again become a force to be reckoned with, and began to exert influence on events abroad, notably in Spain. Still, governments in countries like Britain, France and the USA did not encourage the degree of aircraft development and production that the situation demanded, and it was left to private companies to evolve at their own risk many of the new combat aircraft that were needed. For some countries, their government awoke too late. Poland was crushed by well-equipped German forces in 1939, as were France and other countries in the Spring and early Summer of 1940. The lessons of 1914 had not been remembered.

Perhaps the last word ought to go to the new technique of jet propulsion. Pioneered in both Britain and Germany before the Second World War, it was to thrust the first jet fighters into action during the war and then to spur immense progress in both civil and military aviation, to this day.

Stalwart of the post-war RAF was the de Havilland DH9A, which was used by that service from 1918 to 1931. 'Nine-Acks' fought for the last months of the war on the Western Front, and subsequently in Russia, in their designated role as day-bombers. Thereafter, the DH9A is best remembered as a general purpose aircraft, operating in Iraq and India until replaced by Westland Wapitis. For tropical operations, the aircraft were given an additional radiator and fuel tank, sometimes being provided also with a spare wheel, pilot's survival rations and other gear which was hung all over the airframe, until they looked like flying Christmas trees.

The first capital warship sunk by bombs dropped by American aeroplanes was the ex-German battleship *Ostfriesland* (22 800 tons) during a demonstration of bombing attacks on a stationary, unmanned target at sea by US Army MB-2 bombers, commanded by Brigadier-General William Mitchell on 21 July 1921. During these trials, prior to the sinking of the *Ostfriesland*, US Navy flying-boats had sunk the submarine *U117*, and US Army bombers had sunk the ex-German destroyer *G-102* and the cruiser *Frankfurt* (5100 tons). Subsequently Army bombers sank the battleships *Alabama*, *New Jersey* and *Virginia*. The *Washington*, an unfinished battleship of 32 500 tons, was attacked with and hit by all manner of weapons and was finally sunk by naval gunfire.

DH9A

In the early 1920s, the embryo Japanese naval air arm was trained by a British air mission led by Lord Sempill. This photograph shows a Gloster Sparrowhawk fighter taking off from a platform over a gun turret on HIJMS *Yamashiro*, during training in Tokyo Bay

The first military aircraft designed and built in Czechoslovakia was the Letov S1 reconnaissance and light bombing biplane. It first appeared in 1919/20, and a total of 90 of the S1 and S2 versions were completed. Power was provided by a 260 hp Maybach Mb IVa engine; the SH-1 variant had a maximum speed of 120 mph (193 km/h).

The first American aircraft carrier was the USS *Jupiter*, an ex-collier of 11 050 tons (11 227 tonnes), which was converted to feature a stem-to-stern flight deck of 534 ft (163 m) length in 1920. (Later called *Langley*.)

The first Japanese fighter built for operation from an aircraft carrier was the Mitsubishi 1MF1, designed by the Englishman Herbert Smith. First flown in 1921, the 1MF1 proved very successful during flight trials. In November 1921, **the first Japanese aircraft carrier** was launched, as the *Hosho*, and delivery of 1MF1s began in 1923. Just before this, in February 1923, the first deck take-off and landing in a 1MF1 had

Fokker CV-D of the Swedish Air Force

been made by another Englishman, Captain Jordan. **The first Japanese flyer to take off and land on an aircraft carrier** was First Lieutenant Shun-Ichi Kira of the Imperial Japanese Navy. The 1MF1 was a single-seat biplane, powered by a 300 hp Hispano-Suiza engine and armed with two 7·7 mm machine-guns. Maximum speed was 127 mph (204 km/h).

The first experiments with a pressurised cabin were made on a de Havilland DH4 at Wright Field, Dayton, Ohio, in 1922.

The Vickers Vernon biplane was the first troop transport designed as such from the outset. About 60 were built, in three versions – with two 360 hp Rolls-Royce Eagle VIII, 450 hp Napier Lion II or Lion III engines respectively – first entering service with the RAF in 1922. Twelve troops could be carried, or bombs.

The first air control operation, in which an air force became entirely responsible for the internal security of a nation, began in October 1922. The RAF assumed responsibility for maintaining peace in Iraq, replacing the former large army garrison with two squadrons of Vernon transports, four of DH9As, one of Bristol Fighters and one of Snipe fighters.

The most widely-used inter-war military type produced by the Fokker company, that moved back to Holland in 1918, was the CV. Provision for fitting wings of differing area, and various engines, enabled many combat and observation roles to be performed, and five main variants were produced, designated CV-A to CV-E, of which the last two were the most used. CVs served with many air forces, including those of Finland, Hungary, Italy, The Netherlands, Norway, Sweden and Switzerland. The prototype first flew in 1924 and production CV-Es had a maximum speed of 143 mph (230 km/h). Some CVs were still in service during the early Second World War years.

One of the most successful French aircraft produced between the wars was the Potez 25 reconnaissance and light bombing biplane, first flown in 1925. About 4000 examples were produced, in 87 variants, operated in France and by the forces of some 20 countries. Engines from several French manufacturers were installed, ranging from 450 to 600 hp. Accommodation was for two, and the Potez 25 had a maximum speed of 136 mph (219 km/h) with a 450 hp engine.

The fastest light bomber of the mid-1920s was the Fairey Fox. The prototype flew in 1925, with a 480 hp Curtiss D12 engine, and had a maximum speed of 156 mph (251 km/h). The first of 28 production Foxes were delivered in August 1926 to No 12 Squadron RAF, which has since used a fox's head as its badge. Much faster than other bombers of their time, the

Potez 25

Fairey IIIF of the Fleet Air Arm

General 'Billy' Mitchell

Foxes could out-run then-current fighters. No further production was undertaken for the RAF, but later versions were built in Belgium by Avions Fairey and some of these were used in action against the invading Germans in 1940.

The aircraft operated more than any other type by the Fleet Air Arm during the inter-war period was the Fairey IIIF. The prototype made its first flight on 19 March 1926. Production totalled about 620 aircraft, of which more than

A 100 lb phosphorus bomb from one of the Martin MB-2 bombers under Mitchell's command hits the crowsnest of the target ship *Alabama*

230 were operated by the RAF for general-purpose duties and some 365 by the Fleet Air Arm as spotter-reconnaissance and general-purpose biplanes. The last FAA version was the Mk IIIB, powered by a 570 hp Napier Lion XIA engine and with a maximum speed of 120 mph (193 km/h). Fairey IIIFs were operated from land bases, used as seaplanes and catapulted from naval vessels. The type was pronounced obsolete by the FAA in 1940.

The mainstays of US Army Air Corps bomber units from 1927 to the mid-1930s were Huff-Daland/Keystone twin-engined biplane bombers, built in many small-production versions. All were basically similar externally, differing mainly in engines, tail units and wing configurations. The largest production version was the Keystone B-3A; of which 36 were built, powered by two 525 hp Pratt and Whitney Hornet engines and having a maximum speed of 114 mph (183 km/h). Bomb load was 2500 lb (1134 kg).

Outstanding Soviet heavy bomber in the late 1920s and early 1930s was the twin-engined Tupolev TB-1 (ANT-4). Powered by 680 hp M-17 engines, it carried a crew of six and a bomb-load of 2205 lb (1000 kg). Production of the TB-1 was completed in 1932, by which time about 200 had been built, including some twin-float torpedo-bombers for the Russian Navy. A development of the TB-1 was the TB-3 four-engined bomber, of which 800 were built and served from 1932 to 1941, latterly as paratroop and cargo transports.

The first monoplane medium-bomber to be built for the US Army Air Corps was the Douglas YB-7, a gull-winged aircraft powered by twin 675 hp Curtiss V-1570-27 engines. The undercarriage retracted rearward and open cockpits were provided for the crew of four. Test flown in 1930, only the prototype and seven development aircraft were built for, although proving some 60 mph (96 km/h) faster than the Keystone biplanes then in service, development of all-metal bombers like the Boeing B-9 halted any further production.

The first electrical-mechanical flight simulator was the Link Trainer, which represented a replica of an aeroplane with full controls and instruments, but which did not leave the ground; instead it was 'attached' to a mechanical crab which traced a path over a large-scale map in such a way as to represent heading, speed and

Hawker Fury I fighter exported to Portugal

time of the replica aircraft 'flown' by its occupant. It was invented by Edward Albert Link who sold his first model in 1929; it was adopted by the US Navy in 1931, and by the US Army in 1934. By 1939 there was scarcely an air force in the world that was not using Link Trainers, and there can be no doubt that they were the forerunners of today's complex flight simulators.

The largest Japanese military aircraft built between the wars was the Mitsubishi Ki-20, the basic design of which originated in Germany as the Junkers G 38. The first Ki-20 was completed in secrecy in 1931; by 1934 it had been joined by five others, the later examples built from Japanese components. The Japanese Army Air Force was not impressed by these aircraft, and the Ki-20s never saw action. Each had a wing span of 144 ft 4 in (44 m), was powered by four 800 hp Junkers L88 or 720 hp Junkers Jumo 4 engines and had a range of 1865 miles (3000 km).

The first fighter to enter service with the Royal Air Force capable of a maximum level speed of more than 200 mph (322 km/h) was the Hawker Fury I biplane. Powered by a

525 hp Rolls-Royce Kestrel IIS liquid-cooled engine and armed with two synchronised Vickers machine-guns, the Fury had a top speed of 207 mph (333 km/h) at 14000 ft (4270 m). Designed by the late Sir Sydney Camm, it first entered service with No 43 Squadron, at Tangmere, Sussex, in May 1931.

Two out of every three fighters in service with the French Air Force in 1932 were Nieuport-Delage 62s. Three main versions

P.Z.L. P7as of No 111 Kosciuszkowska Squadron of the 1st Air Regiment, Polish Air Force

Curtiss F9C-2 Sparrowhawks with airship hook-on attachments

were produced – the ND62 powered by a 500 hp Hispano-Suiza 12 Mb engine (345 built), the ND622 with 500 hp H-S 12 Md engine (330 built) and the ND629 with 500 hp H-S 12 Mdsh supercharged engine (50 built). Others were exported. Maximum speed of the initial production version was 150 mph (241 km/h).

The first twin-engined bomber to be designed for the clandestine German Luftwaffe was the Dornier Do 11. Originally designated the Dornier Do F, the prototype first flew on 7 May 1932, disguised as the last of a batch of mail and cargo transport aircraft. Powered by two 650 hp Siemens-Jupiter engines, some 77 Do 11s were produced.

The first monoplane aircraft with retractable landing gear to enter service with the RAF was the Avro Anson general-reconnaissance aircraft.

The first air force in the world to totally equip its front-line fighter units with all-metal monoplanes was the Lotnictwo Wofskowe or Polish Air Force. The aircraft used was the PZL P7a, which featured the famous Pulawski gull-type wing. The prototype P7 first flew in October 1930; altogether some 150 were constructed, deliveries starting in 1932. About 30 were still on strength at the beginning of the Second World War. Powered by a Polish-built 485 hp Skoda Jupiter F VII radial engine, the

Dewoitine D500

Heinkel He 51s of the Luftwaffe

P7a had a maximum speed of 203 mph (326 km/h) and was armed with two Vickers machine-guns.

The first and only American operational fighters to serve aboard airships were naval Curtiss F9C Sparrowhawk biplanes which served on board the US airships *Akron* and *Macon* between 1932 and 1935. The prototype Sparrowhawk (*XF9C-1*) achieved the first 'hook-on' on the airship *Los Angeles* on 27 October 1931, and the first production aircraft hooked-on to *Akron* on 29 June 1932.

The International Disarmament Conference began in Geneva on 2 February 1932. However, Japan left the League of Nations on 27 March 1933 and Germany on 14 October 1933.

The first cantilever low-wing monoplane fighter to go into French Armeé de l'Air service was the Dewoitine D500. The prototype made its first flight on 18 June 1932 but, due to the fact that the wings were found not to be strong enough, delivery did not start until 1935. Variants of the D500 were the D501 and D510. Many examples of the series were exported, the largest foreign operator being China with 34 D510s.

Germany's first new post-war fighters, built to equip the secret Luftwaffe in the early 1930s, were the Heinkel He 51 and Arado Ar 68. Many He 51s were built, in several versions, and 135 were operated by the Spanish and German air forces during the Spanish Civil War. Their useful life in this role was curtailed when the Republican forces started flying Russian Polikarpov I-15s. The He 51B-1 had a maximum speed of 205 mph (330 km/h) and was powered by a 750 hp BMW VI engine.

Grumman FF-1

Farman F222, largest French bomber of the inter-war period

The first prototype Ar 68 initially flew in 1933; the major production version was the Ar 68E which entered service from 1937, although small numbers of Ar 68Fs had entered service in 1935. The Ar 68E had a maximum speed of 208 mph (335 km/h) and was powered by a 690 hp Junkers Jumo 210D engine.

The first multi-engined bombers of the clandestine German Luftwaffe to be used in combat were Junkers Ju 52/3m transports, converted to bombers, with a bucket-like retractable ventral gun position under the fuselage, pending delivery of Dornier Do 17s and Heinkel He 111s. Some served in Spain with the German Legion Cóndor during the civil war.

The first aeroplane flown by the US Navy to have a retractable undercarriage was the Grumman XFF-1 fighter, and as the FF-1 entered service with the Navy in June 1933.

The last operational US biplane to remain in production was the Curtiss SBC Helldiver, which first flew as the parasol monoplane *XF12C-1* in 1933. It was almost immediately redesigned as a biplane, and remained in production in the United States until 1941. At the time of the Japanese attack on Pearl Harbor in December 1941 the US Navy still retained 186 Helldiver biplanes on strength.

The first monoplane fighter with a fully enclosed cockpit and a fully retractable undercarriage to enter squadron service anywhere in the world was the Polikarpov I-16 Ishak ('Little Donkey'). The prototype first flew on 31 December 1933, and deliveries of the Type 1 production fighter to Soviet squadrons commenced during the autumn of 1934. The I-16 Type 1 was powered by a 450 hp M22 engine, had a top speed of 224 mph (360 km/h) at sea-level, and was armed with two 7·62 mm ShKAS machine-guns.

Boulton Paul Overstrand (Charles E. Brown)

Most famous torpedo-bomber/reconnaissance aircraft ever built, the Fairey Swordfish (nicknamed 'Stringbag' because of its profusion of struts and wires) was a 138 mph (222 km/h) biplane in a world of fast low-wing monoplanes. The prototype first flew on 17 April 1934 (as the TSR 2) and production Mk Is entered service with the FAA in July 1935. By the beginning of the Second World War, 13 squadrons were equipped with the type, 12 more being formed during the war. Among its epic victories were the attack on the Italian fleet at Taranto on 11 November 1940, and the crippling of the German battleship *Bismarck*. A total of 2391 Swordfish were built; the Mk I version was powered by a 690 hp Bristol Pegasus III M3 engine. (See also The Second World War.)

The largest French bomber to go into service between the wars was the Farman F221 which, with its derivatives, was the mainstay of the French bomber force for several years. Production began in 1934 with 12 F221s. Later versions included the F222, F222/2, F223 and F223³. The F221 had a wing span of 118 ft 1½ in (36 m) and was powered by four 800 hp Gnome Rhône 14Kbrs engines. Maximum speed was 185 mph (297 km/h).

The first RAF bomber fitted with a power-operated enclosed gun turret was the Boulton Paul Overstrand, 24 of which were built and entered service in 1934. Power was provided by two 580 hp Bristol Pegasus II M3 engines; a bomb load of 1600 lb (725 kg) could be carried.

The Curtiss SOC Seagull served on every aircraft carrier, battleship and cruiser of the US Navy from 1935 to 1945. Designed as a scouting-observation biplane, the first production version was the SOC-1, armed with one

0·30-inch forward-firing Browning machine-gun and a similar gun in the rear cockpit. Two 116 lb bombs could also be carried. Following the original order for 135 SOC-1s, 40 SOC-2s and 130 SOC-3/SON-1s were built. Powered by a 600 hp Pratt and Whitney Wasp engine, the SOC-1 had a maximum speed of 165 mph (265 km/h). It remained in service longer than both types that had been built to replace it.

The existence of the Luftwaffe created by Hitler's Germany was officially announced on 9 March 1935.

The first German monoplane fighter with a fully enclosed cockpit and a fully retractable undercarriage to enter squadron service was the Messerschmitt Bf 109B-1. The prototype Bf 109V-1 (D-IABI) first flew in September 1935 powered by a 695 hp British Rolls-Royce Kestrel V engine. The first production Bf 109B-1s were delivered to Jagdgeschwader 2 'Richthofen' in the spring of 1937, and this version was the first of the series to become operational in Spain with the Legion Cóndor during the civil war. The B-1 model was powered by a 635 hp Junkers Jumo 210D engine, had a top speed of 292 mph (470 km/h) at 13 100 ft (4000 m), and was armed with three 7·92 mm MG 17 machine-guns.

On 3 October 1935 the Italian government declared war on Abyssinia, starting a campaign which was to last until 5 May 1936. The main Italian aircraft used in this conflict were the Caproni Ca 74, Caproni Ca 101, Caproni Ca 111 and Caproni Ca 133 bombers and the Fiat CR20, Fiat CR30 and IMAM Ro 37 fighters.

The Japanese Navy's last carrier-based biplane fighter was the Nakajima A4N1, which had a maximum speed of 219 mph (354 km/h).

Gloster Gladiators of No 1 Squadron, 2nd Regiment, Belgian Air Force

Hawker Hurricane Is of No 111 Squadron, Royal Air Force

Powered by a 770 hp Nakajima Hikari I engine, about 300 examples were built. They first saw combat in 1937, during the Sino-Japanese war, operating from the aircraft carrier *Hosho*.

The first American monoplane fighter with a fully enclosed cockpit and a retractable, though exposed, undercarriage to enter squadron service was the Seversky P-35, the prototype of which

was evaluated at Wright Field in August 1935. The production model, of which deliveries began in July 1937, was powered by a 950 hp Pratt & Whitney R-1830-9 engine. It had a top speed of 281 mph (452 km/h) at 10000 ft (3050 m), and was armed with one 0·5 in and one 0·3 in machine-gun.

The last biplane fighter to serve with the RAF and FAA was the Gloster Gladiator, which entered RAF service from February 1937. As newly-built aircraft, the RAF received 444, while the FAA received 60 Sea Gladiators, serving up to and during the early war years. The Gladiator is probably best remembered for its defence of Malta, when four such fighters fought off large numbers of Italian aircraft.

The first British monoplane fighter with a fully enclosed cockpit and a fully retractable undercarriage to enter squadron service was the Hawker Hurricane. It was also **the RAF's first fighter able to exceed a speed of 300 mph (483 km/h), and the first of its eight-gun monoplane fighters.** The prototype made its first flight on 6 November 1935, and initial deliveries of production aircraft were made to No 111 Squadron at Northolt, Middlesex, during December 1937. The Hurricane I was powered by a 1030 hp Rolls-Royce Merlin II or III engine, had a top speed of 322 mph (518 km/h) at 20000 ft (6100 m), and was armed with eight 0·303 in Browning machine-guns.

The first French monoplane fighter with a fully enclosed cockpit and a fully retractable undercarriage to enter squadron service was the

Consolidated Catalina amphibian (William T Larkins)

Blackburn Skua, the Fleet Air Arm's first monoplane (Charles E Brown)

Morane-Saulnier M-S 406. The M-S 405, from which the series was derived, first flew on 8 August 1935, and was the first French fighter aircraft able to exceed a speed of 250 mph (402 km/h) in level flight. The first production M-S 406 (N2-66) flew for the first time on 29 January 1939, and by 1 April 1939 a total of 27 had been delivered to the French Air Force. Powered by an 850 hp Hispano-Suiza HS 12Y-31 engine, the M-S 406 had a maximum level speed of 304 mph (490 km/h) at 14 700 ft (4480 m), and was armed with one 20 mm HS 59 cannon and two 7·5 mm MAC machine-guns.

The first Japanese Army Air Force low-wing monoplane fighter, and first with an enclosed cockpit, was the Nakajima Ki-27. The first prototype made its initial flight on 15 October 1936, and production aircraft first fought in Manchuria in 1938. Powered by a 710 hp Nakajima Ha 1b engine, the Ki-27 had a fixed undercarriage and flew at a maximum speed of 286 mph (460 km/h). In addition to its two 7·7 mm machine-guns, it could carry 220 lb (100 kg) of bombs.

The flying-boat produced in the greatest numbers was the Consolidated PBY Catalina. Excluding production in Russia, 1196 Catalina flying-boats and 944 amphibians were built, serving with the air forces and airlines of more than 25 nations. Regarded as one of aviation's classic designs, the PBY was still flying in the 1970s in several parts of the world. The prototype first flew in 1935, and the initial production version had a maximum speed of 177 mph (284 km/h).

The first monoplane to enter service with the Fleet Air Arm was the Blackburn Skua

The Macchi C200

dive-bomber, the prototype of which first flew on 9 February 1937. Deliveries started in November 1938 and 165 Skuas were built, remaining in service until 1941.

First monoplane heavy bomber to equip several squadrons of the RAF was the Handley Page Harrow, which entered service with No 214 Squadron in April 1937. Basically interim conversions of a transport aircraft design, 39 Mk Is were built, followed by 61 of the Mk II version with two 925 hp Bristol Pegasus XX engines. Five RAF squadrons were equipped with Harrows at the end of 1937; but from the beginning of the Second World War these aircraft were used as troop transports and minelayers. Weapon load was up to 3000 lb (1360 kg).

The world's first carrier-based monoplane fighter to achieve operational status was Japan's Mitsubishi A5M *Claude* which entered service with the carrier *Kaga* in 1937 during air operations against the Chinese.

The first Italian monoplane fighter with a fully enclosed cockpit and fully retractable undercarriage to enter squadron service was the Macchi C200 Saetta ('Lightning'). First single-seat fighter designed by Dr Mario Castoldi, the prototype first flew on 24 December 1937. Deliveries of production aircraft began in October 1939, and these were powered by an 870 hp Fiat A74RC38 radial engine, giving the C200 a maximum level speed of 313 mph (505 km/h) at 15750 ft (4800 m). Armament consisted of two 12·7 mm Breda-SAFAT machine-guns mounted in the upper engine decking.

The first Japanese monoplane fighter with a fully enclosed cockpit and fully retractable undercarriage to enter squadron service was the Mitsubishi A6M2, popularly known as the 'Zero-Sen'. The A6M1 prototype first flew on 1 April 1939, and the '12-Shi fighter project', as it had been known, was officially adopted by the Imperial Japanese Navy on 31 July 1940. In commemoration of the anniversary of the 2600th Japanese Calendar year (AD 1940), the new fighter became designated A6M2 Type O Carrier Fighter Model 11. The Zero was first used operationally on 19 August 1940, when a formation of 12 aircraft, led by Lieutenant Tamotsu Yokoyama, escorted a force of bombers attacking Chungking. The A6M2 was powered by a 950 hp Nakajima NK1C Sakae 12 engine, had a top speed of 332 mph (534 km/h) at 16570 ft (5050 m), and was armed with two

20 mm Type 99 cannon (licence-built Oerlikons) and two 7·7 mm Type 97 machine-guns. Two 30 kg bombs could be carried on an under-fuselage rack.

The inventor of the first aircraft jet engine to run was Sir Frank Whittle, who was born on 1 June 1907 at Coventry. In 1928, while a Cadet at Cranwell, he published a thesis entitled *Speculation* which put forward the basic equations of thermodynamics for this system. Although the Air Ministry showed interest, the idea was not taken up, and it was not until 1935 that Whittle had some success when R D Williams started a company called Power Jets Ltd to exploit the invention. On 12 April 1937 Whittle ran **the first aircraft turbojet engine in the world** (although a gas-turbine engine had been made previously, he was the first person to make one for aircraft propulsion). The Ministry, in March 1938, gave Whittle a contract for an engine, and on 15 May 1941 a Gloster E28/39, powered by this engine, took off at Cranwell flown by Flight-Lieutenant P E G Sayer. This was the **first British jet aeroplane.** By 1944 Gloster Meteor squadrons were being formed, and flew their first combat sorties against German V-1 flying-bombs.

The Spanish Civil War (1936–9) provided the setting for many significant advances in military aviation. The political background – a rebellion by right-wing elements of the armed forces and population against an extreme left-wing government and popular movement – was such that massive aid was dispatched to the opposing forces by two major international camps. Italy and Germany supported General Francisco Franco y Bahamonde's Nationalists, and the Soviet Union supported the Republican Government, which also drew aid from various other countries and from many volunteer organisations more dedicated in their resistance to Fascism than their outright support of Communism.

German aid, in the form of 20 Junkers Ju 52/3m bomber-transport aircraft, six Heinkel He 51b fighter biplanes and 85 volunteer air and ground crews, arrived in August 1936, less than a month after the outbreak of war. These aircraft were at once employed in ferrying 10 000 Moorish troops across the Straits of Gibraltar from Tetuan. From this small beginning grew the Legion Cóndor – a balanced force of between 40 and 50 fighters, about the same number of multi-engined bombers, and about 100 mis-

Fiat CR32 fighter of the Spanish Nationalist Air Force

cellaneous ground-attack, reconnaissance and liaison aircraft, whose first Commander-in-Chief was General-Major Sperrle. Volunteers from the ranks of the Luftwaffe served in rotation, to ensure the maximum dissemination of combat experience. Many of the major combat designs upon which Germany was to rely in the first half of the Second World War were first evaluated under combat conditions in Spain; the Heinkel He 111 bomber, the Dornier Do 17 reconnaissance-bomber, the Messerschmitt Bf 109 fighter, and the Henschel Hs 123 and Junkers Ju 87 ground-attack aircraft were prominent. The contribution of the Legion Cóndor to the eventual Nationalist victory was considerable, but more important still were the inferences drawn by Luftwaffe Staff planners. Valid lessons learned in Spain included the value of the dive-bomber in hampering enemy communications, and the effects of ground-strafing by fighters in the exploitation of a breakthrough by land forces. Less realistic was the impression gained of the relative invulnerability of unescorted bombers and dive-bombers – an impression based on the lack of sophisticated fighter resistance. In the field of fighter tactics, and in terms of combat experience by her fighter pilots, the Spanish Civil War put Germany at least a year ahead of her international rivals.

Italy's intervention in the Spanish Civil War began in August 1936, with the arrival by sea at Melilla of 12 Fiat CR32 biplane fighters. CR32s were later to become the Nationalists' main fighters, taking over from the slower Heinkel He 51s. The eventual strength of the Italian Aviacion del Tercio in Spain was some 730 aircraft (total supplied) including Fiat CR 32s, SM81s, SM79s, BR20s, Ro37s, Ba65s and a squadron of Fiat G50s. Of these, 86 aircraft were lost on operations and 100 from other

causes, and 175 flying personnel were killed. A total of 903 enemy aircraft were claimed destroyed in aerial combat, and a further 40 on the ground. Total sorties flown were 86420, and bombing sorties totalled 5318.

The Spanish Republican Air Force mustered 214 obsolete aircraft at the outbreak of the Civil War. Additionally, the Government had at its disposal 40 civil types of various designs; and between 1937 and 1939, 55 aircraft were built in the Republican zone. Foreign aircraft dispatched to Spain by various friendly nations totalled 1947, of which 1409 were sent from Russia. The others included 70 Dewoitine D371, D500 and D510 fighters, 20 Loire-Nieuport 46s and 15 S510 fighters from France; 72 aircraft – but no fighters – from the USA; 72 aircraft from the Netherlands; 57 from Britain; and 47 from Czechoslovakia. Of these some 400 are thought to have been destroyed other than in aerial combat, and 1520 were claimed shot down by Nationalist, German and Italian pilots.

The first Russian aircraft to enter combat in Spain were the Polikarpov I-16 Type 6 fighters of General Kamanin's expeditionary command, based at Santander. By September 1936 105 of these aircraft had arrived in Spain – by sea to Cartagena – and had been assembled; some 200 pilots and 2000 other personnel had also arrived from the Soviet Union. The I-16 – known in Spain as the *Rata* ('Rat') and the *Mosca* ('Fly') to Nationalists and Republicans respectively – first entered combat on 5 November 1936. Eventually a total of 475 I-16s were supplied; and from March 1937 they were gathered in one formation designated Fighter Group 31, comprising seven squadrons of 15 aircraft each. More numerous but inevitably less successful was the I-15 biplane fighter, of which some 550 were supplied. The I-15 was inferior to both the Fiat CR32 and the Messerschmitt Bf 109, and no fewer than 415 are believed to have been lost either in combat or on the ground. The most numerous Republican bomber type, the Soviet Tupolev SB-2, also fared badly; of 210 supplied, 178 were lost.

The first production aircraft to embody geodetic criss-cross lattice construction, designed by Barnes Wallis, was the Vickers Wellesley bomber. In 1937, five RAF Wellesleys were modified for an attempt on the world distance record. Modifications included the provision for a third crew member, installation of a

Handley Page Heyford bomber, showing the under-fuselage 'dustbin' turret extended

Pegasus XXII engine in a long-chord cowling, fitting of a constant-speed propeller, and extra fuel tanks. Of the five aircraft delivered to the Long-range Development Flight, RAF, three were chosen for the attempt. Setting off from Ismailia, Egypt, on 5 November 1938, two Wellesleys reached Darwin, Australia, on 7 November, after a record non-stop flight of 7162 miles (11 525 km) in 48 h 5 min flying time. The third aircraft reached its destination a few hours later, after an unexpected fuel stop at Koepang, Dutch East Indies.

The greatest altitude ever achieved by a piston-engined aircraft is 56 046 ft (17 083 m), set by a Caproni 161*bis* on 22 October 1938.

The biggest military aircraft built before the Second World War was the Douglas XB-19. Evaluated by the US Army Air Corps against the Boeing XB-15 (with a 150 ft; 47·72 m wing span), it was powered by four 2200 hp engines and had a maximum speed of 204 mph (327 km/h). The wing span was 212 ft (64·62 m) and it had a weight of 72 tons (73·2 tonnes). Crew numbered 11. Only the prototype was built.

The fastest aircraft up to the Second World War was a special development of the German Messerschmitt Bf 109 single-seat fighter. Bf 109s in Luftwaffe service at the outbreak of war included B, C, D and E series aircraft, the last with a maximum speed of 354 mph (570 km/h). However, on 26 April 1939, a much-changed Me 209 flew at 469 mph (755 km/h) at Augsburg, Germany, setting a record that was not beaten by another piston-engined aeroplane until 30 years later.

The first flight by a turbojet-powered aircraft took place on 27 August 1939, when the Heinkel He 178, piloted by Flugkapitan Erich Warsitz, flew for the first time at Marienhe airfield. Power plant was the Heinkel HeS 3b engine designed by Dr Pabst von Ohain.

The last biplane heavy bomber to serve as such with the Royal Air Force was the Handley Page Heyford which first flew in June 1930 at Radlett, Hertfordshire, entered service with No 99 (Bomber) Squadron at Upper Heyford, Oxfordshire, on 14 November 1933, and was withdrawn from front-line service in March 1939. The last Heyford was removed from the RAF inventory in May 1941. Powered by two 525 hp Rolls-Royce Kestrel III engines, the Heyford I had a top speed of 142 mph (229 km/h) at 13 000 ft (3960 m) and carried a normal bomb-load of 2800 lb (1270 kg).

The aircraft carrier limitations imposed by the Washington Treaty (signed on 6 February 1922) permitted the United States and Great Britain to build aircraft carriers up to a total tonnage of 135 000 tons (137 170 tonnes) per country; Japan could build 81 000 tons (82 300 tonnes), and France and Italy, 60 000 tons (61 000 tonnes) each. Aircraft carriers were not permitted to carry guns of greater calibre than 8 in (20 cm). Moreover, carriers already under construction at the time of the signing of the Treaty were exempt from its provisions. A limit of 27 000 tons (27 500 tonnes) was placed on each ship, but each nation might possess up to two vessels each of 33 000 tons (33 500 tonnes). The USS *Saratoga* and *Lexington* each displaced approximately 33 000 tons (33 500 tonnes), but at full load the figure was nearer 40 000 (40 650). When no longer considered to be bound by the provisions of the Treaty, Japan rebuilt several of her carriers during the 1930s to displace more than 33 000 tons (33 500 tonnes). Of all the signatory nations (as major Sea Powers) of the Washington Treaty, only Italy chose to ignore the significance of the aircraft carrier, and constructed no such ship between the two world wars.

AIRCRAFT CARRIERS COMPLETED BETWEEN THE WORLD WARS

Name	Displacement at full load tons	Length ft (m)	Speed kt	Aircraft accommodation	Completed	Ultimate fate	Remarks
GREAT BRITAIN							
Furious	22 450	786 (239·6)	32·5	33	Recommissioned 1925	Scrapped in 1949	Converted from light battle-cruiser during construction. Originally completed in 1917.
Argus	15 775	565 (172·2)	20·76	20	1918	Scrapped in 1947	Converted from liner *Conte Rosso* during construction.
Eagle	26 400	667 (203·3)	24	21	1920	Sunk by U-Boat, 1942	Converted from battleship *Almirante Cochrane* during construction. The first aircraft carrier in the world with offset 'island' superstructure and split-level aircraft hangar. Originally laid down as a battleship for Chile, in 1913; purchased and completed as a carrier in 1920.
Hermes	12 900	598 (182·3)	25	25	1923	Sunk by Jap a/c, 1942	World's first carrier designed as such. Laid down in January 1918 and completed in July 1923, it was the first carrier in the world to be sunk by aircraft from another carrier.
Courageous	26 500	786 (239·6)	32	48	1928	Sunk by U-Boat, 1939	Converted from light battle-cruiser.
Glorious	26 500	786 (239·6)	32	48	1930	Sunk by German warships, 1940	Converted from light battle-cruiser.
Ark Royal	27 000	800 (243·8)	31·5	60	1938	Sunk by U-Boat, 1941	Max accommodation 72 aircraft.
UNITED STATES OF AMERICA							
Langley	11 050	542 (165·2)	15	34	1922	Sunk by Jap a/c, 1942	Converted from collier *Jupiter*. The first American aircraft carrier commissioned for fleet service; completed in September 1922.
Lexington	40 000	888 (270·7)	34	80	1927	Sunk by Jap a/c, 1942	Converted from battle-cruiser during construction. Max accommodation 120 aircraft.
Saratoga	40 000	888 (270·7)	34	80	1927	Destroyed in A-Bomb test, Bikini, 1946	Remarks as for *Lexington*.
Ranger	14 500	769 (234·4)	29·5	80	1934	Scrapped in 1947	
Yorktown	19 900	741 (225·9)	29·5	80	1937	Sunk by Jap a/c, 1942	
Enterprise	19 900	741 (225·9)	29·5	80	1938	Scrapped in 1958	

Name	Displacement at full load tons	Length ft	m	Speed kt	Aircraft accommodation	Completed	Ultimate fate	Remarks
JAPAN								
Hosho	7470	551	(167·9)	25	21	1922	Scrapped in 1947	First Japanese aircraft carrier to be laid down as such, in December 1919; completed in December 1922.
Akagi	36500	855	(260·6)	31·2–32·5	60	1927	Sunk by US a/c, 1942	Converted from battle-cruiser during construction. Accommodation increased to 90 a/c in 1938 on rebuilding.
Kaga	38200	812	(247·5)	27·5–28·3	60	1930	Sunk by US a/c, 1942	Converted from battleship during construction. Aircraft accommodation increased to 90 in 1936.
Ryujo	10600	590	(179·8)	29	48	1933	Sunk by US a/c, 1942	
Soryu	15900	746	(227·4)	34·5	73 (max)	1937	Sunk by US a/c, 1942	
Hiryu	17300	746	(227·4)	34·3	73 (max)	1939	Sunk by US a/c, 1942	
FRANCE								
Bearn	25000	599	(182·6)	21·5	40	1927	Scrapped in 1968	First French aircraft carrier. Converted from a Normandie-class battleship during construction. Completed in 1927. Not used as aircraft carrier (sic) after SecondWW but used as an aircraft transport. The Bearn was **the longest-lived carrier in the world.**

At the outbreak of the Second World War there were 20 aircraft carriers in commission and 11 under construction by the world's major naval powers:

	Completed	Under construction
Great Britain	7	6
Japan	6	2
United States of America	6	2
France	1	1

Triumphant procession of the de Havilland DH 50J, which made the famous survey flights to Capetown and Australia in the mid-1920s, flown by Alan Cobham.

SECTION 6

BETWEEN THE WARS - CIVIL

For nearly a decade after the First World War, the United States concentrated on carrying mail, rather than passengers, by air. As a result, airlines got off to a late start in the nation where powered flying had begun in 1903. The **first mail flights between New York and Washington** began while the war was still being fought, on 15 May 1918, using Curtiss JN-4H 'Jennies' regarded as no longer good enough for military training duties. The old biplanes covered the 218 mile (350 km) route at only 50–60 mph (80–96 km/h). The time was improved when the US Post Office took over the service, on 12 August 1918, using specially-built Standard biplanes. But the aircraft of that period were too slow to offer any significant advantage over surface travel on such

a short route, and the New York–Washington service closed on 31 May 1921.

The first flight from Egypt to India was made by Captain Ross M Smith, DFC, AFC, Major-General W G H Salmond, DSO, Brigadier-General A E Borton, and two mechanics between 29 November and 12 December 1918, in a Handley Page O/400 from Heliopolis to Karachi via Damascus, Baghdad, Bushire, Bandar Abbas and Chabar. The same aircraft had made **the first flight from England to Egypt** between 28 July and 8 August 1918.

The first sustained commercial daily passenger service was by Deutsche Luft-Reederei, which operated between Berlin and Weimar,

Germany, from 5 February 1919. Aircraft used on the service were five-seat AEG biplanes and two-seat DFWs. The 120 mile (193 km) flight took 2 h 18 min.

The first airline passenger flight between Paris and London was made on 8 February 1919 by a Farman F60 Goliath, owned by the Farman brothers and piloted by Lucien Bossoutrot. As civil flying was not yet permitted in the UK, the token payload consisted of military passengers, who flew from Toussus le Noble to Kenley. The flight is not recognised as a genuine scheduled operation.

The first purely commercial aircraft to be built for passenger carrying in Britain after the First World War, and one of the first new civil types in the world, was the de Havilland DH16, the prototype of which flew for the first time in March 1919. The type entered service with Aircraft Transport and Travel Ltd in May. Altogether, nine DH16s were built by June 1920; the first six were powered by a 320 hp Rolls-Royce Eagle VIII engine; the others had a 450 hp Napier Lion engine. Accommodation was for four passengers. The last DH16 was withdrawn from use in August 1923.

The first sustained (but not daily) regular international service for commercial passengers was opened between Paris and Brussels by the Farman brothers on 22 March 1919. Fare for the 2 h 50 min flight was 365 francs. The pilot was again Lucien Bossoutrot. The **first Customs examination of passengers** took place at Brussels after the third of the weekly flights, on 6 April.

The first American international air mail was inaugurated between Seattle, Wash, and Victoria, British Columbia, Canada, by the Hubbard Air Service on 3 March 1919 using a Boeing Type C aircraft. The service was regularised by contract on 14 October 1920.

Civil flying in Britain restarted, after the First World War, on 1 May 1919, following publication of the Air Navigation Regulations.

The first stage of a planned US transcontinental air mail service was inaugurated on 15 May 1919, between Chicago and Cleveland. Realising that only a really long air route would show any advantage over surface transportation in terms of journey time, the US Post Office had purchased more than 125 ex-military aircraft, mostly de Havilland DH4Ms, with which it hoped to carry air mail between New York and San Francisco at the standard first class surface rate of 2 cents an ounce. The final link from Omaha to Sacramento, over the Rockies, was proved practicable on 8 September 1920. After months of careful planning, **the first coast-to-coast air mail in the USA** left San Francisco at 04.30 h on 22 February 1921 and arrived at Mineola, Long Island, New York, at 16.50 h the following day. It was carried from San Francisco to North Platte, Nebraska, by a succession of pilots. At North Platte, it was taken over by a pilot named Jack Knight, flying one of the open-cockpit DH4Ms. When he reached Omaha, weather conditions along the route were so bad that the pilot scheduled to fly the next stage to Chicago had not put in an appearance. So Knight carried on through the darkness, over unfamiliar country, to Chicago, becoming a national hero for 'saving the mail service'. A system of lighting, for safer flying at night, was installed along part of the route in 1923. Regular transcontinental night mail flights were inaugurated on 1 July 1924.

The first occasion on which newspapers were distributed by an aircraft on a daily basis was in May 1919. A Fairey IIIC seaplane was used on a week-long experimental freighting service to distribute the *Evening News* from the Thames, near Westminster Bridge, to coastal towns in Kent. This service saved about two hours' distribution time.

The first British aeroplane to carry civil markings (*K-100*) was a de Havilland DH6 in 1919. It was sold to the Marconi Wireless Telegraph Co Ltd, and used for radio trials; it became the second aircraft entered on the British Civil Register (as *G-EAAB*; see below).

The first British civil aeroplane (i.e. the first on the British Civil Register proper) was a de Havilland DH9 (*G-EAAA*, previously *C6054*), operated as a mailplane by Aircraft Transport and Travel Ltd, in mid 1919 between London and Paris.

The first Transatlantic crossing by air was achieved by the American Navy/Curtiss NC-4 flying-boat commanded by Lieutenant-Commander A C Read between 8 May and 31 May 1919. Three flying-boats, the *NC-1*, *NC-3* and *NC-4*, under the command of Commander John H Towers, set out from Rockaway, NY, on 8 May, but only *NC-4* completed the crossing, arriving at Plymouth, England, on 31 May, having landed at Chatham, Mass; Halifax, Nova

Farman F60 Goliath piloted by Lucien Bossoutrot

Scotia; Trepassey Bay, Newfoundland; Horta, Azores; Ponta Delgada, Azores; Lisbon, Portugal; Ferrol del Caudillo, Spain. Total distance flown was 3925 miles (6315 km) in 57 flying h, 16 min, at a speed of 68 mph (110 km/h) Both *NC-1* and *NC-3* were forced down on the sea short of the Azores, and *NC-1* sank – its crew being rescued. *NC-3*, with Commander Towers aboard, taxied the remaining 200 miles (320 km) to the Azores. The crew of the *NC-4* were:
Lieutenant-Commander A C Read
  (Commander)
Lieutenant E F Stone (pilot)
Lieutenant Walter Hinton (pilot)
Ensign H C Rodd (radio operator)
Lieutenant James L Breese (engineering officer)
Chief Machinist's Mate E S Rhoades (engineer)

The first regular civil air service in England was started by A V Roe and Co on 10 May 1919 and was discontinued on 30 September 1919. Using three-seat Avro aircraft, services were flown from Alexander Park, Manchester, to Southport, and also to Blackpool. One hundred and ninety-four scheduled flights were made during this period; the cost of a one-way flight was four guineas.

The first non-stop air crossing of the Atlantic was achieved on 14–15 June 1919 by Captain John Alcock and Lieutenant Arthur Whitten Brown who flew in a Vickers Vimy bomber from St John's, Newfoundland, to Clifden, County Galway, Ireland. Powered by two Rolls-Royce engines, the Vimy was fitted with long-range fuel tanks and achieved a coast-to-coast time of 15 h 57 min. Total flying time was 16 h 27 min. Both Alcock and Brown were knighted in recognition of this achievement; Sir John Alcock, as Chief Test Pilot of Vickers, was killed on 18 December 1919 in a flying accident in bad weather near Rouen, France.

The first post-war airline to operate an airship was DELAG, a subsidiary of the Zeppelin Company, which began regular flights from

The AEG J II used on the first Berlin–Weimar services, 1919

NC-4, the first aircraft to fly across the Atlantic, in stages

Friedrichshafen to Berlin on 24 August 1919, ending operations on 1 December of the same year. The airship *Bodensee* began the service and, altogether, about 103 flights were made, which carried some 2400 passengers and 66 140 lb (30 000 kg) of cargo. Each flight carried 23 passengers.

The first Dutch national airline, KLM (Royal Dutch Airlines), was founded on 7 October 1919. It is the oldest airline in the world still operating under its original name.

The first airline to provide food on its services was Handley Page Transport, in October 1919, when it introduced lunch baskets.

The first scheduled daily international commercial airline flight anywhere in the world was flown from London to Paris on 25 August 1919 when a de Havilland DH16, flown by Cyril Patteson of Aircraft Transport and Travel Ltd., took off from Hounslow with four passengers, and landed at Le Bourget, Paris, 2.5 h later. The fare was £21 for the one-way crossing.

The first American international scheduled passenger air service was inaugurated on 1 November 1919 by Aeromarine West Indies Airways between Key West, Florida, and Havana, Cuba. On 1 November 1920 the airline was awarded the first American foreign air-mail contract.

The first flight from Britain to Australia was completed between 12 November and 10 December 1919 by two Australian brothers, Captain Ross Smith and Lieutenant Keith Smith, with two other crew members, in a Vickers Vimy bomber powered by two Rolls-Royce Eagle engines. They set out from Hounslow, Middlesex, England, and flew to Darwin, Australia, a distance of 11 294 miles (18 175 km) in under 28 days. Their feat earned them the Australian Government's prize of £10 000 ($40 000) and knighthoods. Sir Ross Smith was killed in a flying accident near Brooklands Aerodrome, England, on 13 April 1922. By tragic coincidence, both Sir Ross and Sir John Alcock

Alcock and Brown preparing for take-off, 14 June 1919

(famous for his Atlantic crossing, see above) were killed in Vickers Viking amphibians.

The first successful large British passenger aircraft, designed as such, was the Handley Page W8, the prototype of which first flew on 4 December 1919. Following the success of converted Handley Page O/400 bombers as civil passenger aircraft, Handley Page Transport operated three W8bs on its London to Paris airline service from May 1922. Powered by two 360 hp Rolls-Royce Eagle VIII engines, the W8b had a maximum speed of 104 mph (167 km/h) and could accommodate 12 passengers.

The first flight across Australia was made during the period 16 November–12 December 1919 by a BE2e flown by Captain H N Wrigley, DFC, accompanied by Lieutenant A W Murphy, DFC, from Melbourne to Darwin to meet Ross and Keith Smith. They covered the 2500 miles (4000 km) in 46 h flying time.

The first flight from Britain to South Africa was made between 4 February and 20 March 1920 by Lieutenant-Colonel Pierre van Ryneveld and Squadron Leader Christopher Quintin Brand. They set out from Brooklands, England, in a Vickers Vimy bomber on the 4 February, but crashed at Wadi Halfa while attempting an emergency landing. The South African Government provided the pilots with another Vimy aircraft and after 11 days they set off again, only to crash at Bulawayo, Southern Rhodesia, on 6 March. Once again the Government provided an aircraft, a war-surplus DH9, and on 17 March they set off again. Finally, on 20 March, they reached Wynberg Aerodrome, Cape Town. They received subsequently £5000 prize-money and were knighted by HM King George V.

The first regular use of Croydon as London's air terminal was on 29 March 1920. On that day the main airport facilities were moved from Hounslow (see above) to Croydon – or Waddon, as the airport was originally known. It was opened officially on 31 March 1921.

The first automatic pilot to be fitted in a British commercial aircraft was the Aveline Stabiliser fitted in a Handley Page O/10 during 1920.

The oldest airline service operated still by the original airline company is the London–Amsterdam service which was first flown by KLM (Royal Dutch Airlines) on 17 May 1920.

Handley Page W8

Aircraft used for the pioneer flight was an Eagle-engined DH16 (G-EALU) of Aircraft Transport and Travel, piloted by H 'Jerry' Shaw. After that, all early KLM services were operated by this British airline under charter.

The first Australian commercial airline, QANTAS (Queensland and Northern Territory Aerial Service) was registered on 16 November 1920 for air taxi and regular air services in Australia. The Company's first Chairman was Sir Fergus McMaster (1879–1950), and its first scheduled service commenced on 2 November 1922 with flights between Charleville and Cloncurry, Queensland.

The first fatal accident to a scheduled British commercial flight occurred on 14 December 1920 when a Handley Page O/400 crashed soon after take-off in fog at Cricklewood, London. The pilot, R Bager, his engineer and two passengers were killed, but four other passengers escaped.

The first regular scheduled air services in Australia were inaugurated by West Australian Airways on 5 December 1921.

The first air collision between airliners on scheduled flights occurred on 7 April 1922 between a Daimler Airways de Havilland DH18 (G-EAWO) flown by Robin Duke from Croydon, and a Farman Goliath of Grands Express Aériens flown by M Mier from Le Bourget. The two aircraft, which were following a road on a reciprocal course, collided over Thieuloy-Saint-Antoine 18 miles (29 km) north of Beauvais. All seven occupants were killed.

The first coast-to-coast crossing of the United States in a single day was made by Lieutenant James H Doolittle who flew a modified de Havilland DH4B from Pablo Beach,

Florida, to Rockwell Field, San Diego, Calif, on 4 September 1922. Actual flying time to cover the 2163 miles (3480 km) was 21 h 19 min; elapsed time, with a refuelling stop at Kelly Field, Texas, was 22 h 35 min.

The first German aeroplane to land in the UK post-war was a Dornier Komet four-passenger airliner (D-223) of Deutsche Luft-Reederei, which visited Lympne aerodrome, in Kent, on 31 December 1922.

The first scheduled air service between London and Berlin was inaugurated by Daimler Airway on 10 April 1923, with intermediate landings at Bremen and Hamburg.

The first non-stop air crossing of the United States of America by an aeroplane was achieved on 2–3 May 1923 by Lieutenant O G Kelly and Lieutenant J A Macready of the US Air Service in a Fokker T-2 aeroplane. Taking off from Roosevelt Field, Long Island, at 12.36 h (Eastern Time) on 2 May, they arrived at Rockwell Field, San Diego, Calif at 12.26 h (Pacific Time) on 3 May. They overflew Dayton, Ohio; Indianapolis, Ind; St Louis, Mo; Kansas City, Mo; Tucumcari, New Mexico; and Wickenburg, Ariz. The distance flown, 2516 miles (4050 km), was covered in 26 h 50 min. Kelly and Macready also established a new world's endurance record for aeroplanes on 16–17 April 1923 in the Fokker T-2, by flying a distance of 2518 miles (4052 km) over a measured course in 36 h 5 min.

The first successful in-flight refuelling of an aeroplane was accomplished by Captain L H Smith and Lieutenant J P Richter in a de Havilland DH4B on 27 June 1923 at San Diego, Calif, USA. Smith and Richter established a world's endurance record by remaining aloft for 37 h 15 min 43·8 s during 27–28 August 1923, covering a distance of 3293·26 miles (5299·9 km) over

DH18

a measured 50 km course at San Diego, Calif. Their DH4B was flight-refuelled 15 times.

The first British national airline, Imperial Airways, was formed on 1 April 1924. This was the manifestation of the British Government's determination to develop air transport, and the company was to receive preferential air subsidies, having acquired the businesses of the British Marine Air Navigation Co, Daimler Airway, Handley Page Transport and Instone Air Lines. Its fleet consisted of seven DH34s, two Sea Eagles, three Handley Page W8bs and one Vickers Vimy.

The first successful round-the-world flight was accomplished by two Douglas DWCs (Douglas World Cruisers) between 6 April (leaving at 08.47 h) and 28 September 1924 (landing at 13.30 h). Four DWCs set out from Seattle, Washington, and two of them circumnavigated the world. The flagplane (named *Seattle* and piloted by the commander of the flight, Major F Martin) only reached Alaska, where it crashed on a mountain on 30 April; the crew returned to the USA. Another (the *Boston*) was forced to alight in the ocean near The Faeroes, and eventually sank after being cut loose from the crew's rescue ship, the USS *Richmond*. The crew of the two successful flights were as follows:
Chicago (Aircraft No 2): Lieutenant Lowell H Smith, deputy flight commander and pilot; Lieutenant Leslie P Arnold, mechanic and alternate pilot.
New Orleans (Aircraft No 4): Lieutenant Erik Nelson, pilot and engineer officer; Lieutenant John Harding, Jr, mechanic and maintenance officer.

Starting point was Seattle, Washington; the journey ending with flights from Washington, DC, to Eugene, Oregon, and then back to Seattle.

The total mileage flown by each of the two aeroplanes (Nos 2 and 4) which completed the historic flight was 27 553 miles (44 340 km) and the elapsed time 175 days. The flying time was given as 371 h 11 min.
(Note on the Douglas DWC amphibians. These aircraft were specially ordered by the US Army as adaptations of the Douglas DT Navy torpedo bomber. Powered by a single in-line engine, the twin-float biplanes had a span of 50 ft (15·24 m) and a fuel capacity of 600 US gal (2271 litres). Two of the World Cruisers are preserved to this day, one at the

Smithsonian Institution in Washington, the other in the Air Force Museum at Wright-Patterson Air Force Base.)

The first three-engined all-metal monoplane transport in the world to enter commercial airline service was the Junkers G 23, four of which served with Swedish Air Lines (AB Aerotransport) on the Malmö–Hamburg–Amsterdam route, commencing on 15 May 1925. This aircraft, built in Germany and Sweden, provided the basis for the later, famous Junkers Ju 52/3m, and set the pattern for low-wing multi-engined monoplanes thereafter.

The first return flight between London and Cape Town was made by Alan Cobham, with his mechanic A B Elliott and ciné photographer B W G Emmott, in the second de Havilland DH50 that was built (G-EBFO). They left Croydon on 16 November 1925 and reached Cape Town on 17 February 1926. Cobham landed back at Croydon after the 16 000 mile (25 750 km) round trip on 13 March 1926. Three months later, on 30 June, Cobham and Elliott took off from the River Medway in the same aircraft, converted into a seaplane for a largely overwater flight to Australia and back. G-EBFO returned to a triumphant welcome on the Thames, at Westminster, on 1 October. The success was marred by the death of Elliott, who had been killed by a stray bullet fired by a bedouin while flying over the desert between Baghdad and Bazra on the outward leg. Cobham was knighted for these flights, which pioneered future Empire air routes.

The first aeroplane flight over the North Pole was accomplished by Lieutenant-Commander Richard E Byrd (USN), and Floyd Bennett in a Fokker F.VIIA/3m called *Josephine Ford* on 9 May 1926. The total distance flown was 1600 miles (2575 km).

The first scheduled and sustained airline service to carry passengers in the USA was that operated by Western Air Express, between Salt Lake City and Los Angeles, from 23 May 1926.

The first successful passenger aircraft built in America was the Ford 4-AT Trimotor, which first flew in June 1926. Powered by three 220 hp engines, it could accommodate ten passengers and had a cruising speed of 105 mph (169 km/h). From 1926 to 1933, Ford built around 200 Trimotors, designated from 3-AT to 7-AT, and the type was operated by many airlines.

A duration record was set up at Tirlement, Belgium by Crooy and Groenen who kept the lower of these two DH9s airborne for 60 h 7 min 32 s by using the flight refuelling technique pioneered by Smith and Richter (Shell)

Two great airliners of the mid 1920s were the de Havilland DH66 Hercules and the Armstrong Whitworth Argosy. The Hercules was a 79 ft 6 in (24·23 m) span biplane, powered by three 420 hp Bristol Jupiter VI engines. It was designed to Imperial Airways requirements, to fulfil the need for a large multi-engined transport suitable for tropical operations, when the airline agreed to take over the air mail services from Cairo to Baghdad that had been pioneered by the RAF. The prototype first flew on 30 September 1926 and, following minor modifications, was sent to Heliopolis on 18 December. Named *City of Cairo*, the DH66 made its first commercial flight on 12 January 1927. Altogether, 11 DH66s were built, seven for service with Imperial Airways and four for West Australian Airways. The last to survive was *City of*

The Douglas World Cruisers, lined up for take-off at the start of their round-the-world flight

Adelaide, which was destroyed in New Guinea in 1942. The British aircraft normally accommodated a crew of three, seven passengers and mail. The Australian aircraft accommodated 14 passengers. **The Armstrong Whitworth Argosy,** a 20-passenger airliner, was powered by three 420 hp Jaguar engines. It was the Argosy that flew Imperial Airways' lunchtime 'Silver-Wing' service to Paris.

The first non-stop solo air crossing of the Atlantic was made by Captain Charles Lindbergh during 20–21 May 1927 in a single-engine Ryan high-wing monoplane, *Spirit of St Louis*, from Long Island, New York, to Paris, France. (It is perhaps worth emphasising that this crossing was being made at the same time as Flt Lt C R Carr was attempting to fly non-stop from England to India, and there is little doubt that had the latter not been forced to alight in the Persian Gulf, Lindbergh's epic flight might well have been eclipsed to some extent.) It took Lindbergh 33 h 39 min to complete the 3610 mile (5810 km) journey, at an average speed of 107·5 mph (173 km/h).

The first non-stop air crossing of the South Atlantic by an aeroplane was made on 14–15 October 1927 when Captain Dieudonné Costes and Lieutenant-Commander Joseph Le Brix flew a Breguet XIX aircraft, *Nungesser-Coli*, from Saint-Louis, Senegal, to Port Natal, Brazil, a distance of 2125 miles (3420 km) in 19 h 50 min.

The first air service operated by Pan American Airways was inaugurated on 19 October 1927 on the 90 mile (145 km) route between Key West, Florida, and Havana, Cuba.

The first solo flight from Great Britain to Australia was made by Squadron-Leader H J L ('Bert') Hinkler in the Avro 581 Avian prototype light aircraft, *G-EBOV*, flying from Croydon, London, to Darwin, Australia, between 7 and 22 February 1928. His 11005 mile (17711 km) route was via Rome, Malta, Tobruk, Ramleh, Basra, Jask, Karachi, Cawnpore, Calcutta, Rangoon, Victoria Point, Singapore, Bandoeng and Bima. His aircraft was placed on permanent exhibition in the Brisbane Museum.

The Australian Flying Doctor Service was inaugurated on 15 May 1928 using the joint services of the Australian Inland Mission and QANTAS at Cloncurry. The first aircraft was a de Havilland DH50 *Victory*, modified to accommodate two stretchers; its first pilot was A Affleck and the first flying doctor was Dr K H Vincent Welsh. The founder of the service was the Reverend J Flynn, OBE.

The first trans-Pacific flight was made between 31 May and 9 June 1928 by the Fokker F.VIIB-3m *Southern Cross* flown by Captain Charles Kingsford Smith and C T P Ulm (pilots), accompanied by Harry Lyon (navigator) and James Warner (radio operator), from Oakland Field, San Francisco, Calif, to Eagle Farm, Bris-

First flight over the N. Pole was accomplished in the Fokker F.VIIA/3m trimotor named *Josephine Ford*

bane, via Honolulu, Hawaii, and Suva, Fiji. The flight covered 7389 miles (11890 km), with a flying time of 83 h 38 min. The *Southern Cross* has been preserved and is displayed at Eagle Farm Airport. It was **the first aircraft ever to land in Fiji.**

The first large-scale airlift evacuation of civilians in the world was undertaken by transport aircraft of the Royal Air Force between 23 December 1928 and 25 February 1929 from the town of Kabul, Afghanistan, during inter-tribal disturbances. Five hundred and eighty-six people and 24193 lb (10975 kg) of luggage were airlifted over treacherous country using eight Vickers Victoria transports of No 70 Squadron, RAF, and a Handley Page Hinaidi.

The first commercial air route between London and India was opened by Imperial Airways on 30 March 1929. The route was from London to Basle, Switzerland, by air (Armstrong Whitworth Argosy aircraft); Basle to Genoa, Italy, by train; Genoa to Alexandria, Egypt, by air (Short Calcutta flying-boats); Alexandria to Karachi, India, by air (de Havilland DH66 Hercules aircraft). The total journey from Croydon to Karachi occupied seven days, for which the single fare was £130. The stage travelled by train was necessary as Italy forbade the air entry of British aircraft, an embargo which lasted several years and substantially frustrated Imperial Airways' efforts to develop the Far East route.

The first non-stop flight from Great Britain to India was accomplished by Squadron Leader A G Jones Williams, MC, and Flight-Lieutenant N H Jenkins, OBE, DFC, DSM (pilot and navigator respectively) between 24 and 26 April 1929. Flying from Cranwell, Lincolnshire, to Karachi, India, in a Fairey Long Range Monoplane, *J9479*, powered by a 530 hp Napier Lion engine, they covered the 4130 miles (6647 km) in 50 h 37 min. It had been intended to fly to Bangalore to establish a world distance record but the attempt was abandoned owing to headwinds. However, another Fairey Long Range Monoplane, *K1991*, was crewed by Squadron Leader O R Gayford and Flight-Lieutenant G E Nicholetts on **the first non-stop flight from England to South Africa,** between 6 and 8 February 1933. The total distance covered from Cranwell, Lincolnshire, to Walvis Bay, South-West Africa, was 5309 miles (8544 km), which was completed in 57 h 25 min and set a new world record.

Charles Lindbergh with the Ryan monoplane *Spirit of St Louis*

The largest flying-boat built between the wars was the Dornier Do X, the prototype of which made its maiden flight on 25 July 1929, at which time it was also **the largest aeroplane of any kind in the world.** Powered by 12 engines (initially 525 hp Siemens Jupiters, later 600 hp Curtiss Conquerors), mounted in tandem pairs on the 157 ft 5¾ in (48 m) monoplane wing, the Do X once carried (on 21 October 1929) a crew of ten and 159 passengers (nine of which were stowaways). Altogether three Do Xs were built. The most famous flight was that made by the prototype from Germany to New York, which took from 2 November 1930 to 27 August 1931 because of damage to the wing and hull and lengthy stops at various places including Amsterdam, Calshot, Lisbon, the Canary Islands, Portuguese Guinea and Rio de Janeiro. In commercial terms the Do X was quite impracticable, and the two examples delivered to an Italian airline were never put into service. Although it had a respectable maxi-

DH50 of Qantas used on the original flying doctor service, 1928

Vickers Victoria of No 70 Sqn RAF at Risalpur after flight from Kabul, 1929

mum speed of 134 mph (216 km/h), the Do X weighed 123 460 lb (56 000 kg) with full load and had a service ceiling of only 1640 ft (500 m).

The first airship flight round the world was accomplished by the German Graf Zeppelin between 8 and 29 August 1929. Captained by Dr Hugo Eckener, the craft set out from Lakehurst, New Jersey, and flew via Friedrichshafen, Germany, Tokyo, Japan, and Los Angeles, Calif, returning to Lakehurst 21 d, 7 h 34 min later.

The two large British commercial airships, R-100 and R-101, were completed at the end of 1929, the R-101 (G-FAAW) flying first at Cardington, Bedford, on 14 October, and the R-100 (G-FAAV) on 16 December, flying from Howden, Yorkshire, to the Royal Airship Works, Cardington. The R-100 made a transatlantic flight from Britain to Canada between 29 July and 1 August 1930, returning between 13 and 16 August. The R-101 was destroyed on 5 October 1930 when it struck a hill near Beauvais, France, during an intended flight from Cardington to

India. Commanded by Flight-Lieutenant H C Irwin, AFC, the R-101 was carrying 54 people, of whom 48 were killed or died later, including Lord Thompson, Secretary of State for Air, and Major-General Sir Sefton Brancker, Director of Civil Aviation. This tragedy ended British efforts to develop airships for commercial use.

The first flight over the South Pole was made by Commander R E Byrd, US Navy,

Armstrong Whitworth Argosy *City of Glasgow* (Charles E Brown)

Fairey Long Range Monoplane K1991

with Bernt Balchen (pilot), Harold June (radio) and Ashley McKinley (survey), during 28–29 November 1929, in a Ford 4-AT Trimotor monoplane, named *Floyd Bennett*.

The first solo flight from Great Britain to Australia by a woman was achieved by Miss Amy Johnson between 5 and 14 May 1930, flying a de Havilland DH60G Gipsy Moth *Jason* (*G-AAAH*) from Croydon to Darwin.

The first airline stewardess was Ellen Church, a nurse who, with Boeing Air Transport (later absorbed into United Air Lines), made her first flight between San Francisco, Calif, and Cheyenne, Wyoming, on 15 May 1930.

The first coast-to-coast all-air commercial passenger service in America was opened by Transcontinental and Western Air Inc. between New York and Los Angeles, Calif, on 25 October 1930.

The first flight of the Handley Page 42E Hannibal four-engine biplane airliner, *G-AAGX*, was made on 14 November 1930, as a prototype, from Radlett, Hertfordshire. The stately HP 42/45 class brought new standards of luxury and reliability to air travel besides providing the backbone of Imperial Airways' fleet during the 1930s. Eight aeroplanes were produced in two versions, the 'E' and 'W', which could carry 24 and 38 passengers respectively.

The first commercial air route between London and Central Africa was opened on 28 February 1931 by Imperial Airways. The route lay from Croydon to Alexandria (using Argosy aircraft from Croydon to Athens, and Calcutta flying-boats from Athens to Alexandria via Crete), and from Cairo to Mwanza, on

Dornier Do X photographed in the UK, where passengers on local flights included the Prince of Wales (later King Edward VIII)

Ford 4-AT Trimotor *Floyd Bennett*

Lake Victoria (using Argosy aircraft). Passengers were only carried as far as Khartoum, mail being carried over the remainder of the route.

On 26 March 1931 the Swiss airline companies Balair and Ad Astra joined forces to form Schweizerische Luftverkehr, or Swissair. The initial fleet consisted primarily of six Fokker F VIIbs; but in 1932, a fast new service was inaugurated using Lockheed Orions. The Orion, which was the first commercial aircraft to be fitted successfully with a retractable under-carriage, was a cantilever low-wing monoplane of wooden construction, powered by a 420 hp Pratt and Whitney Wasp engine. Able to fly at 225 mph (362 km/h), it made Swissair the fastest operator on the Zürich, Munich and Vienna air routes.

Between 23 June and 1 July 1931 a Lockheed Vega named *Winnie Mae*, piloted by Wiley Post and with Harold Gatty as navigator, completed a round-the-world flight in a record 8 d 15 h 51 min, starting and finishing at New York. The flight covered a distance of nearly 15 500 miles (24 945 km) in about 106 flying hours, following a route via Chester (England), Berlin, Irkutsk (Russia) and Alaska (see also page 113).

Imperial Airways' first monoplane airliner was the Armstrong Whitworth AW15 Atalanta, eight of which were constructed for that airline in 1931 and 1932. Accommodation was for nine passengers plus cargo or mail, and the aircraft were used on the Nairobi–Cape Town and Karachi–Singapore routes.

Handley Page 42 *Horatius* at Croydon Airport

Winnie Mae, the Lockheed Vega flown by Wiley Post

Amelia Earhart with her Lockheed Vega

The first formation flight across the South Atlantic, from Portuguese Guinea to Natal, Brazil, was made by ten Savoia-Marchetti S55 flying-boats, commanded by Italian Air Minister General Italo Balbo, on 6 January 1931. In 1933, from 1–15 July, General Balbo led the first formation flight across the North Atlantic. Taking off from Orbetello, Italy, the 24 S55s flew to the Century of Progress Exposition in Chicago, their route being via Holland, Iceland, Labrador and New Brunswick, Canada.

The first solo crossing of the North Atlantic by a woman was made by the American pilot Miss Amelia Earhart (Mrs Putnam) during 20–21 May 1932, when she flew a Lockheed Vega aircraft from Harbour Grace, Newfoundland, to Londonderry, Northern Ireland.

The true ancestor of modern airliners, with all-metal structure, cantilever low wing and a retractable undercarriage, was the Boeing Model 247, of which 75 were built. The prototype first flew on 8 February 1933; the major production version was the Model 247D, with two 550 hp

Pratt and Whitney Wasp engines driving controllable-pitch propellers. Cruising speed was 189 mph (304 km/h), with ten passengers and 400 lb (181 kg) of mail.

The first flights over Mount Everest were made on 3 April 1933 by the Marquess of Clydesdale, flying a Westland PV-3, and by Flight-Lieutenant D F McIntyre flying a Westland Wallace, each with one passenger.

The Douglas DC-1 first flew on 1 July 1933 at Clover Field, Santa Monica, Calif. Only one was built, flying mainly in TWA markings. After service during the Spanish Civil War, it was written off in a take-off accident near Malaga, Spain, in December 1940. The DC-1 never killed a passenger. From it were developed the famous DC-2 and DC-3, which established the reputation of Douglas as an airliner manufacturer.

Savoia-Marchetti S55s of General Balbo's 1933 transatlantic formation flight

The first solo flight round the world was achieved by Wiley Post between 15 and 22 July 1933, when he flew his Lockheed Vega monoplane, *Winnie Mae*, from Floyd Bennett Field, New York, for a distance of 15 596 miles (25 099 km) in 7 d 18 h 49 min. His route was via Berlin, Moscow, Irkutsk and Alaska, back to New York. Post later pioneered the development of an early pressure suit during high-altitude flights in *Winnie Mae*. He was killed in a crash in 1935, together with the famous comedian/philosopher Will Rogers.

The first regular internal air mail in Great Britain was carried by Highland Airways on 29 May 1934. The de Havilland Dragon *G-ACCE*, flown by E E Fresson, carried 6000 letters from Inverness to Kirkwall, and the service operated thereafter on every week-day.

Westland PV-3 approaching Everest, 3 April 1933

The first flight by an aeroplane from Australia to the United States was made by a Lockheed Altair aircraft flown by Sir Charles Kingsford Smith accompanied by Captain P G Taylor from Brisbane to Oakland, Calif, via Fiji and Hawaii, between 22 October and 4 November 1934.

The first regular weekly air-mail service between Britain and Australia commenced on 8 December 1934 from London to Brisbane via Karachi and Singapore. The airlines participating were Imperial Airways, Indian Trans-Continental Airways and Qantas Empire Airways. Mail which left London on this day reached Brisbane on 21 December.

A privately-sponsored civil aircraft that led to one of Britain's main bomber types of the early war years was the Bristol 142. Ordered by Lord Rothermere of the *Daily Mail* as a six-passenger high-speed transport aircraft, at a cost of around £18 500, the 142 first flew on 12 April 1935. Powered by two 650 hp Bristol Mercury engines, it proved to have a maximum speed of 307 mph (494 km/h). This was some 80 mph (128 km/h) faster than the RAF's latest fighter, the Gloster Gauntlet. On seeing that the Air Ministry was suitably impressed with his aircraft, Lord Rothermere gave the 142 to the nation and named it *Britain First*. A bomber derivative was soon on the production line as the Bristol Blenheim.

The first through passenger air service between London and Brisbane, Australia, was inaugurated on 13 April 1935 by Imperial Airways and Qantas Empire Airways. The single fare for the 12 754 mile (20 525 km) route was £195. However, owing to heavy stage bookings no through passengers were carried on the inaugural flight. The journey took 12½ days.

On 1 October 1935 British Airways was formed from Hillman Airways, Spartan Air Lines and United Airways. It was to be merged with Imperial Airways in 1940, to form British Overseas Airways Corporation (BOAC).

The first solo crossing of the South Atlantic by a woman was accomplished by New Zealand's Jean Batten, flying a Percival Gull from Lympne, Kent, to Natal, Brazil, via Thies, Senegal during the period 11–13 November 1935. Eighteen months earlier, between 6 and 28 May 1934, Jean Batten had flown solo from Lympne to Darwin, Australia, in a de Havilland

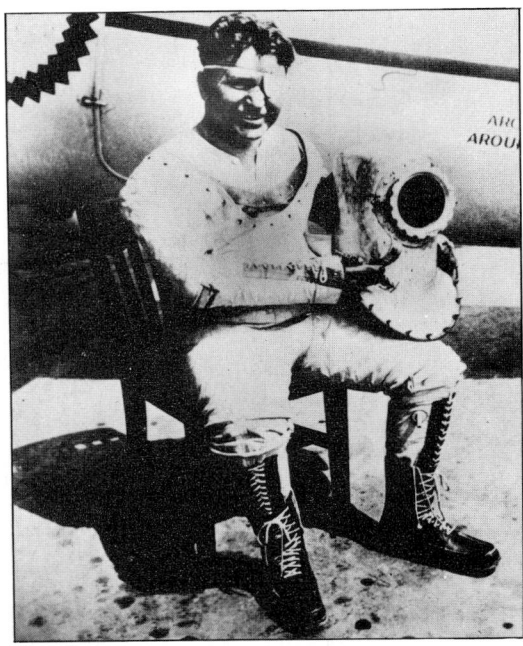

Wiley Post in pioneer pressure suit

Moth. Between 5 and 16 October 1936, she flew solo from Lympne to her native New Zealand, receiving a heroine's welcome at Auckland.

On 22 November 1935 and 21 October 1936, Pan American Airways inaugurated its first trans-Pacific mail service and its trans-Pacific passenger service respectively. The route was between San Francisco and Manila in the Philippines, via Honolulu, Midway Island, Wake Island and Guam, and took about six days. The aircraft used was the Martin 130 China Clipper, which was a high-wing monoplane flying-boat, powered by four 830 hp Pratt and Whitney Twin Wasp engines. Maximum cruising speed was 163 mph (262 km/h); range was 3200 miles (5150 km), and accommodation was for up to 43 passengers by day or 18 in a night sleeper layout.

The first flight of the Douglas DC-3 prototype was made on 17 December 1935, flown by Carl A Cover from Clover Field, Santa Monica. A development of the DC-1 (Douglas Commercial No 1) and DC-2, it first entered service with American Airlines and its inaugural passenger-carrying service was from Chicago, Illinois

Bristol 142 *Britain First*

to Glendale, Calif on 4 July 1936. Certainly the most famous airliner in aviation history, large numbers remain in civil and military service in 1977, 42 years after the prototype's first flight.

The first case in Britain involving the prosecution and punishment of an airline passenger for smoking on board an aircraft in flight was heard at Croydon, Surrey, on 17 March 1936. The passenger, who had been on an Imperial Airways Paris–London flight aboard the HP45 *Heracles*, was fined the sum of £10.

The first British Airways began using Gatwick Airport as its operating base on 17 May 1936.

The first Short C-Type Empire flying-boat *Canopus* (*G-ADHL*) made its first flight on 4 July 1936 with John Lankester Parker, Short's Chief Test Pilot, at the controls. Its first flight with Imperial Airways was made on 30 October 1936. The Empire boats represented the last word in luxury travel before the Second World War and, as their name implied, were flown on the Empire routes to Africa and the Far East. The Empire flying-boat also represented **one of the biggest gambles in commercial aviation history** as 28 such aircraft were ordered by Imperial Airways before the first aircraft was built.

The first direct, solo east-to-west crossing of the North Atlantic by a woman was made by Mrs Beryl Markham during 4–5 September 1936, flying from Abingdon, England, to Baleine, Nova Scotia, where she crashed after a flight of 24 h 40 min. She was not hurt.

The first aircraft with a completely successful pressurised cabin was the Lockheed XC-35 of 1937, built for research at high altitude.

Martin 130 *China Clipper*

Between the Wars continued on page 118

SPAD VII 245 Vieux Charles, *flown by Capitaine Georges Guynemer of Escadrille SPA 3, Groupe de Combat No 12 'Les Cigognes', Service Aéronautique Française. Guynemer, whose fifty-four confirmed victories place him second on France's roll of aces, flew this aircraft early in 1917; full details of his career may be found in the text. The stork insignia was the badge of SPA 3.*

Nieuport 17 1895, *flown by Lieutenant Charles Nungesser of Escadrille N65 during the summer of 1917. Third among French aces with forty-five confirmed victories, Nungesser was a pilot of enormous determination in the face of constant pain; his career is described in the text. He first used his macabre personal marking in November 1915, and from that time onwards it was painted on all his aircraft. From May 1917 onwards the wings (and later the fuselage top decking) of his aircraft were painted with broad tricolor stripes as an additional identification; as in that month he accidentally shot down a British aircraft which attacked him.*

Nieuport 17 *flown during the summer of 1917 by Sergeant Marius Ambrogi of Escadrille N90. 'Marc' Ambrogi was a specialist in shooting down heavily defended observation balloons; these hazardous targets accounted for ten of his fourteen confirmed victories. He survived the war, and later rejoined the colours to fight in the Second World War; in 1940 he was serving as Deputy Commander of Groupe de Chasse I/8 when he shot down a Junkers Ju 52/3m.*

Nieuport 28 *flown by Lieutenant Douglas Campbell of the 94th Aero Squadron, American Expeditionary Force, in March 1918. On 19 March he accompanied Major Raoul Lufbery and Lieutenant Edward Rickenbacker on the first patrol over enemy lines by an American unit; and on 14 April he became the first American-trained pilot to score an aerial victory, his victim being an Albatros shot down near the squadron's base at Toul. His fifth victory on 31 May 1918 made him the first ace who had served exclusively with the American forces. He was to score one further victory before a wound put him out of the war on 6 June 1918; he was eventually discharged in 1919 with the rank of Captain.*

SPAD XIII 1620, *flown by Lieutenant David E. Putnam of the 139th Aero Squadron, American Expeditionary Force, in the summer of 1918. Putnam was officially credited with twelve victories, but unofficial estimates indicate a considerably higher total. He flew with various French units between December 1917 and July 1918 and gained his first victory on 19 January 1918. On 5 June he fought an epic battle against ten enemy aircraft, sending five down destroyed or damaged. He was transferred in July to the 139th Squadron, in which he served as a Flight Commander until his death in action on 12 September 1918.*

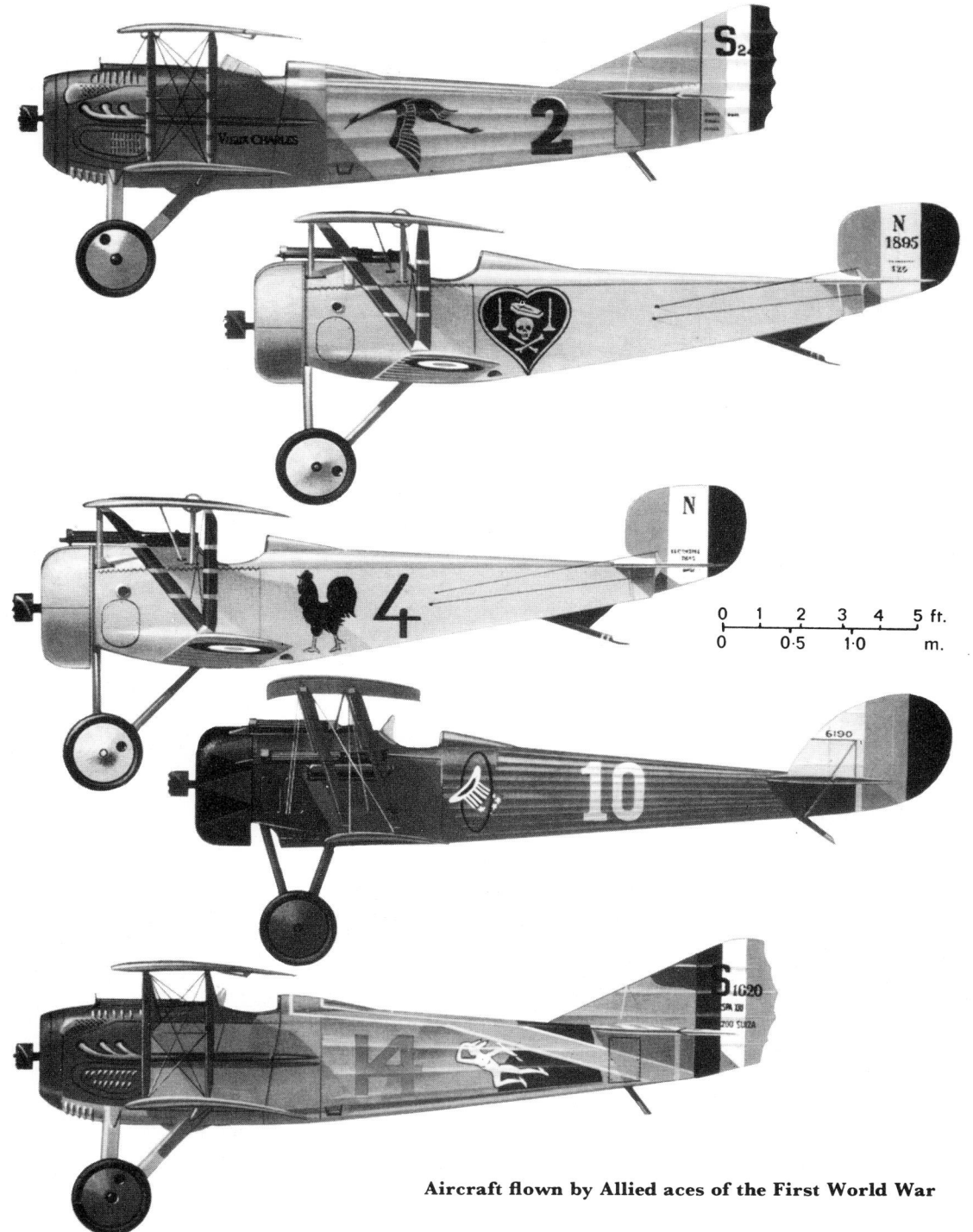

Aircraft flown by Allied aces of the First World War

Short 'C' Class flying-boat *Challenger* of Imperial Airways (Norman Sinclair)

The first commercial survey flights over the North Atlantic were carried out simultaneously by Imperial Airways and Pan American Airways during 5–6 July 1937. The former flew the long-range C-Class flying-boat *Caledonia G-ADHM* westwards from Foynes, Ireland, to Botwood, Newfoundland, while the latter flew the Sikorsky S-42 *Clipper III* eastwards.

The first commercial use of composite aeroplanes in the world occurred during 21–22 July 1938 when the Short S21 *Maia* flying-boat and the Short S20 *Mercury* seaplane took off from Foynes, Ireland, the upper component then separating and flying the Atlantic non-stop to Montreal, Canada, with a load of mail and newspapers. It covered 2930 miles (4715 km) in 20 h 20 min, at an average speed of 140 mph (225 km/h). Numerous composite flights and separations were carried out and the pair of aircraft continued to operate on the Southampton to Alexandria air route until the outbreak of the Second World War. When launched from its 'mother-plane' *Maia*, the seaplane *Mercury* carried sufficient petrol to fly 5997½ miles (9652 km) from Dundee, Scotland, to the Orange River, South Africa. In doing so, on 6–8 October 1938, it set up a record that has never been beaten.

The first flight of a Danish airliner to the United Kingdom was made on 28 July 1938, the inaugural flight being made by the Focke-Wulf 200 Condor *Dania* (*OY-DAM*) between Copenhagen and Croydon.

The largest airliner built for Imperial Airways between the wars was the Armstrong Whitworth Ensign, which entered service on the London–Paris route on 26 October 1938. Powered originally by four 850 hp Tiger engines, the Ensign was later re-engined with four

Heinkel He 178, the first jet aeroplane to fly

950 hp Wright Cyclones. Accommodation was for 27–40 passengers, depending on the area of operation.

The first airliner to enter service with a pressurised cabin was the Boeing Model 307 Stratoliner, the prototype of which first flew on 31 December 1938. Derived from the B-17 Flying Fortress bomber, it was advertised as the first airliner to fly above most bad weather conditions because of its pressurised cabin. The first version to go into service was the Model 307B, with Trans-Continental and Western Air in April 1940. Accommodation was for 33 passengers.

The first regular air mail service over the North Atlantic began on 20 May 1939 with the departure of the Pan American Airways' Boeing 314 *Yankee Clipper* (NC18603) from New York.

The first aircraft in the world to fly solely on the power of a turbojet engine was the German Heinkel He 178, which made its first true flight at Heinkel's Marienehe airfield on 27 August 1939.

The first transatlantic flight by a British Prime Minister was made by the Rt Hon Winston Churchill on 16–17 January 1942. The flight, between Bermuda and Plymouth, was made in the Boeing 314 flying-boat *Berwick* operated by British Overseas Airways Corporation (BOAC).

The Short-Mayo Composite

Douglas TBD-1 Devastators

SECTION 7

THE SECOND WORLD WAR

Looked at in retrospect, it is amazing to discover how unprepared were the combatant nations at the beginning of the Second World War. Of the European nations involved initially, only Germany could be said to have made methodical efforts to assemble a modern air force. The Luftwaffe's participation in the Spanish Civil War had done much to improve combat techniques, and the new tactic of using dive-bombers to provide close support for mobile armoured units on the ground was the beginning of the *Blitzkrieg* concept which, in the early stages of the war, swept all before it.

Many thousands of miles remote from the shores of the continental United States, the Japanese Army and Navy had been steadily building up air forces which, when unleashed across the face of the Pacific, following the initial attack on Pearl Harbor on 7 December 1941, were to come as a shock to the Western Allies. The extent to which war with China had encouraged the development of new, much improved aircraft had not been appreciated. Furthermore, military commanders of the West had failed to understand the extent to which fanatical devotion to their home islands had made the Japanese airmen formidable adversaries, even when they lacked adequate training later in the war.

Before the war's end, there was an unprecedented degree of development over the entire range of aviation equipment. Piston-engines and propellers were victims at the beginning of a power plant revolution, with the introduction of the first crude, but immensely promising, turbojet engines. Radar, which at the war's beginning had appeared to show promise for the detection and location of enemy aircraft, was developed for airborne interception of intruders and gave completely new standards of bombing accuracy. Machine-guns began to lose pride of place to new, more potent cannon; while rocket projectiles gave to medium-size fighter aircraft the 'broadside of a battleship'.

Startling new weapons emerged, including the pilotless V1 'flying bomb' from Germany, as well as the V2 ballistic rocket against which there was no defence other than the elimination of its launching sites. These, and early guided bombs, gave a hint of the possibilities that could – and would – be evolved as new techniques of control and miniaturisation were developed.

The end of the war against Japan was hastened by use of the world's first atomic bombs, dropped over the cities of Hiroshima and Nagasaki. Man had acquired a new, crude weapon which, when refined and wedded to the ballistic rocket, seemed to offer an ultimate weapon. So potent were such weapons, capable of destroying our civilised world, that their awesome ability was to become, perhaps, our best insurance for continuing, if costly, peace.

The last Curtiss biplane to serve in an operational role with the US Navy was the SOC Seagull, a scout–observation aircraft for operation from battleships and cruisers. First flown in April 1934, the Seagull was a floatplane which was catapulted from its parent vessel and had to alight on the sea for recovery. Later versions were modified for carrier operations with wheel landing gear and arrester gear. It remained in first-line service until the end of the Second World War, enduring far longer than two types of aircraft intended to replace it.

The most famous naval aircraft in British service, and in first-line operational use throughout the whole of the Second World War, first flew as a seaplane on 10 November 1934. Fairey Swordfish made the epic attack on

Lockheed Hudsons of RAF Coastal Command
(Charles E Brown)

the Italian fleet, anchored in Taranto harbour, on the night of 11 November 1940. This was the first occasion on which aircraft had virtually eliminated an enemy fleet. Equipped with ASV radar, a Swordfish of No 812 Squadron, based at Gibraltar, became **the first aircraft to destroy a U-boat (U-451) by night,** on the night of 21–2 December 1941. (See also Between the Wars - Military.)

The first operational American naval aircraft to feature hydraulically-operated folding wings was the Douglas TBD Devastator, which was also the US Navy's first carrier-based monoplane torpedo-bomber to enter production. First flight of the prototype was made on 15 April 1935, but it was not until 5 October 1937 that the first production TBD-1s were delivered to Squadron VT-3. Of 75 Devastators on strength with the US Navy on 4 June 1942 – the day of the Battle of Midway – 37 were lost during the course of the battle, Squadron VT-8 being destroyed in its entirety and another squadron decimated in combat with Japanese Zero fighters. Following this action the type was withdrawn from operational use.

The first of the four-engined Boeing bomber aircraft to enter service, initially with the USAAC, was the B-17 Flying Fortress. Powered by four Wright R-1820 engines, the name Flying Fortress was a registered copyright of the Boeing company, chosen to emphasise the heavy defensive armament of this aircraft. The first twelve Y1B–17s entered service during early 1937, equipping the 2nd Bombardment Group. The most heavily armed version of the Fortresses was the B–17G, which carried thirteen 0·5 in machine-guns. This was also the most extensively built version, with 8680 aircraft coming from three production lines. By the war's end, production of the Flying Fortress totalled 12731 aircraft.

Unique among USAAF fighter aircraft was the Bell P–39 Airacobra, the XP–39 prototype of which flew for the first time in April 1938, with an Allison inline engine mounted in the fuselage aft of the pilot. The conventional tractor propeller was driven via a long and nose-mounted reduction gear, this unusual configuration being adopted to allow for the installation of a 37 mm nose-mounted cannon. Because the engine was installed more or less on the aircraft's centre of gravity, a tricycle landing gear was essential, making the Airacobra **the first tricycle-gear single-engined fighter ordered by the United States Army.**

The first United States single-seat fighter aircraft to be built on a mass production basis, the Curtiss P–40 Warhawk, was evaluated by the USAAF in 1939. The initial $13 million order for 524 aircraft was the largest order ever placed for a US fighter aircraft up to that time.

The most famous Japanese naval aircraft of the Second World War was the Mitsubishi A6M Navy Type O Carrier Fighter, a designation giving rise to its identification as the Zero. The first operational use of Zeros was in China; their first major operation was against Pearl Harbor, and adjacent Philippine bases of the US forces, on 7 December 1941. (See also Between the Wars – Military.)

Junkers Ju 87, remembered as the 'Stuka'

Curtiss P-40 Warhawk in the Solomon Islands, August 1943 (USAF)

One of the classic German fighter aircraft of the Second World War, the Focke-Wulf Fw 190, flew for the first time on 1 June 1939. Entering service with the Luftwaffe in August 1941, Fw 190s were engaged in combat with Spitfires for the first time on 27 September 1941. Their first major deployment was to provide an air umbrella for the German battleships *Gneisenau* and *Scharnhorst* and the cruiser *Prinz Eugen* during their 'Channel dash' on 12/13 February 1942.

The first American-built aircraft to be used operationally by the RAF during the Second World War was the Lockheed Hudson, which entered service with No 224 Squadron at Gosport in the Summer of 1939. Hudsons shot down the first German aircraft in the Second World War, directed Naval forces to the prison ship *Altmark*, and took part in the hunting of the battleship *Bismarck*. On 27 August 1941 a Hudson of No 269 Squadron caused submarine U–570 to surrender to it: **the first U-boat captured in a solely RAF operation.**

The greatest losses suffered by the Luftwaffe during the Polish campaign in a single day were those of 3 September 1939 when 22 German aircraft were destroyed (four Dornier Do 17s, three Messerschmitt Bf 110s, two Heinkel He 111s, three Junkers Ju 87s, two Messerschmitt Bf 109s, three Henschel Hs 126s, two Fieseler Fi 156s, one Henschel Hs 123, one Junkers Ju 52 and one Heinkel He 59). One of the Messerschmitt Bf 110s was accidentally shot down by German troops near Ostrolenka. Luftwaffe personnel casualties on this day amounted to 34 killed, one wounded and 17 missing.

The first British aircraft to attack a German U-boat during the Second World War was an Avro Anson I of the RAF's No 500 Squadron then based at Detling. On 5 September 1939, this aircraft made a bombing attack on the enemy submarine.

The first German aircraft to be shot down by British aircraft during the Second World War was a Messerschmitt Bf 109E destroyed by Sgt. F Letchford, air gunner of Fairey Battle number K9243 of No. 88 Squadron, Advanced Air Striking Force of the RAF, over France, on 20 September 1939. The **first Fleet Air Arm victory** followed when three Dornier Do 18s of *Küstenfliegergruppe 506* were sighted by a patrol of Swordfish aircraft flying from HMS *Ark Royal* over the North Sea on 26 September 1939. Nine Skuas were forthwith launched from the carrier and these succeeded in forcing one of the Dorniers (Werke Nr 731, of 2 *Staffel*, Kü Fl Gr 506) down on to the sea in German Grid Square 3440. The German four-man crew was later rescued and made prisoner on board a British destroyer.

The first German aircraft shot down by an RAF aircraft operating from the United Kingdom during the Second World War was a Dornier Do 18 flying-boat destroyed by a Lockheed Hudson of No 224 Squadron. This occurred on 8 October 1939 during a patrol by the RAF aircraft over Jutland.

The first German aircraft shot down over British soil in the Second World War was a Junkers Ju 88A-1 of I/KG30, piloted by Hauptmann Pohl, and destroyed by a Spitfire of No 603 Squadron on 16 October 1939, over the Firth of Forth.

The most famous Japanese bomber aircraft of the Second World War, one of the outstanding aircraft of that war, and the most extensively-built Japanese bomber was the Mitsubishi G4M Navy Type 1 Attack Bomber. First flown on 23 October 1939, the G4M – known to the Allies as *Betty* – was first used operationally in May 1941 in an attack on Chungking. In service throughout the entire Pacific War, the last operational flight of two G4Ms was to carry the Japanese surrender delegation to Ie-Shima on 19 August 1945.

The first enemy aircraft shot down by RAF fighters on the Western Front in the Second World War was a Dornier Do 17 destroyed over Toul on 30 October 1939. The victorious

Consolidated B-24 Liberator (USAF)

pilot was Pilot Officer P W Mould, flying a Hurricane of No 1 Squadron.

The first long-range anti-shipping squadron of the German Luftwaffe was formed in November 1939. In the absence of a more suitable aircraft, the unit was equipped with the Focke-Wulf Fw 200 Condor long-range civil transport. One of these aircraft (D-2600) *Immelmann III* was Adolf Hitler's personal transport. Though not ideal for military service, the Fw 200 was responsible for the destruction of an immense number of Allied merchant ships. The unit flying Condors I/KG40, controlled by the German Navy, claimed more than 363 000 tons (368 800 tonnes) of shipping destroyed during one six-month period.

The first four-engined attack bomber to be designed for the Japanese Navy, and the first aircraft with a retractable tricycle-type landing gear to be built in Japan, was the Nakajima G5N Shinzan which flew for the first time in December 1939. Designed and built after examination of the American Douglas DC-4E prototype, it was found to have indifferent performance and only six were built.

The most extensively-built American aircraft of the Second World War, supplied to the USAAF, USN and Allied air forces, was the Consolidated B-24 Liberator, the prototype of which flew for the first time on 29 December 1939. This four-engined bomber, of which more than 18000 examples were built, was recognisable easily on the ground by virtue of its tricycle landing gear, and in the air by its large twin endplate fins and rudders. On 1 August 1943, 177 Liberators were used unescorted in an heroic attack against the oil refineries at Ploesti in Romania, from bases in North Africa. It involved a 4345 km (2700 mile) round trip during which 57 of the attacking force were lost.

The first Royal Air Force aircraft to drop bombs deliberately on German soil is believed to have been an Armstrong Whitworth Whitley, *N1380*, DY-R of No 102 Squadron, based at Driffield, Yorkshire. No 102 Squadron, in company with Whitleys of Nos 10, 51 and 77 Squadrons, and Handley Page Hampdens of No 5 Group, attacked the German mine-laying seaplane base at Hornum on the night of 19/20 March 1940.

The worst losses in aircraft destroyed as the result of air combat and anti-aircraft gunfire suffered by a single air force on a single day are believed to have been those of the Luftwaffe on 10 May 1940. On this day Germany invaded the Netherlands and Belgium and was opposed simultaneously by the air forces of Holland, Belgium, France and Great Britain. The Norwegian campaign, by then nearing its end, also claimed a small number of German victims. On this day the Luftwaffe, according to its own records, lost:

Junkers Ju 52 transports	157 destroyed
Heinkel He 111 bombers	51 destroyed, 21 damaged
Dornier Do 17 bombers	26 destroyed, 7 damaged
Fieseler Fi 156 artillery support aircraft	22 destroyed
Junkers Ju 88 bombers	18 destroyed, 2 damaged
Junkers Ju 87 dive-bombers	9 destroyed
Messerschmitt Bf 109 fighters	6 destroyed, 11 damaged
Dornier Do 215 reconnaissance aircraft	2 destroyed
Henschel Hs 126 reconnaissance aircraft	1 destroyed, 3 damaged
Messerschmitt Bf 110 fighters	1 destroyed, 3 damaged
Dornier Do 18 flying-boat	1 destroyed
Henschel Hs 123 dive-bomber	1 damaged
Other types	10 destroyed, 3 damaged
Total	304 destroyed, 51 damaged

Aircrew casualties amounted to 267 killed, 133 wounded and 340 missing; other Luftwaffe personnel (Flak, engineers, etc) casualties amounted to 326 killed or missing.

Apart from the purely academic significance of these figures, they indicate conclusively that the operations undertaken by the Luftwaffe on this day represented the true commencement of *Blitzkrieg* against substantial opposition. On this day Germany suffered losses in excess of all previous cumulative losses since 1 September 1939, including the Polish campaign. Losses suffered by the Luftwaffe during the invasion of Poland may be summarised as follows:

1–8 September 1939:
116 aircraft destroyed, 128 aircrew killed, 68 wounded and 137 missing.
9–13 September 1939:
34 aircraft destroyed, 15 aircrew killed, 15 wounded and 63 missing.
14–18 September 1939:
23 aircraft destroyed, 24 aircrew killed, 32 wounded and 14 missing.
19–27 September 1939:
30 aircraft destroyed, 54 aircrew killed, 18 wounded and 4 missing.

The greatest single victory achieved by the Royal Netherlands Air Force during the German invasion of the Low Countries was gained at 06.45 h on 10 May 1940 when a force of Fokker D XXIs intercepted 55 Junkers Ju 52/3m transport aircraft of KGzbV9. The Dutch pilots claimed to have shot down 37 of the formation, but German records indicate a total loss of 39 aircraft, 6 occupants killed, 41 presumed dead, 15 wounded and 79 missing.

The French military aeroplane most widely used at the beginning of the Battle of France on 10 May 1940 belonged to the Potez 630 series, of which a total of 1250 were built. The main variants were the 630 and 631 fighters and the 63/II reconnaissance machine. First flown in April 1936, the three-seater Potez 630 fighter, powered by two 640 hp Hispano-Suiza HS 14AB 10/11 engines, had a maximum level speed of 280 mph (450 km/h) at 13 000 ft (3960 m).

Fokker D XXIs of the Royal Netherlands Air Force in war markings. No 241 took part in a battle between 9 D XXIs and 9 Messerschmitt Bf 109s on 10 May 1940. Four or five of the Bf 109s were shot down for the loss of one D XXI – No 241

Potez 63/II

Standard armament comprised two nose-mounted Hispano 9 or 404 cannon, plus one MAC machine-gun for rear defence. Shortage of cannon made it necessary to arm many 630/631s with four machine-guns and when, in February 1940, it was decided to increase the fire-power of these fighters, the cannon were supplemented by six machine-guns mounted beneath the wings.

The American fighter of the Second World War to remain in production longest was the Vought F4U Corsair which first flew on 29 May 1940, and at which time it was the most powerful naval fighter in the world. Initial deliveries to VF-12 (US Navy Fighter Squadron Twelve) began on 3 October 1942, the type remaining in production until December 1952. It was thus **the last piston-engined American fighter to remain in production.** Fastest of the series was the F4U-5N, powered by a 2300 hp Pratt and Whitney R-2800-32W radial engine with a two-stage supercharger, which gave a maximum level speed of 470 mph (756 km/h) at 26 800 ft (8168 m). In service with the British Fleet Air Arm on board HMS *Victorious*, Corsair Mk IIs took part in the attacks on the German battleship *Tirpitz* on 3 April 1944, this being the first operation flown with Corsairs from an aircraft carrier.

The first bombing attack on Berlin was made on the night of 25/26 August 1940. The attacking force comprised 12 Hampdens of Nos 61 and 144 Squadrons, 17 Wellingtons of Nos 99 and 149 Squadrons, and 14 Whitleys of Nos 51 and 78 Squadrons.

The highest-scoring Allied pilot during the Battle of Britain was Sergeant Josef Franti-šek, a Czech pilot who served with No 303 (Polish) Squadron, RAF. His confirmed score of 17 enemy aircraft shot down was achieved entirely during September 1940; he was killed on 9 October 1940. The only British gallantry decoration awarded to František was the Distinguished Flying Medal, but he had previously been awarded the Czech War Cross and the Polish Virtuti Militari.

The first Victoria Cross to be won during the Battle of Britain was awarded posthumously to Acting Seaman J F (Jack) Mantle, RN, who was operating an anti-aircraft gun aboard HMS *Foyle Bank* in Portland Harbour, Dorset, on 4 July 1940. The ship, the only one in port with an anti-aircraft gun, became the focus of an enemy raid and was hit by a bomb which cut the power-supply. Jack Mantle, though severely wounded, continued to fire the gun, operating it manually; despite another direct hit upon the ship, which severed his left leg, he remained at his post until the end of the raid but succumbed to his terrible wounds almost immediately afterwards. His Victoria Cross was only the second to be awarded for an action in or over Great Britain, the first having been awarded to Lieutenant W Leefe Robinson of No 39 (Home Defence) Squadron, RFC, for his destruction of a Schutte-Lanz airship on the night of 2/3 September 1916 at Cuffley, Hertfordshire.

The first Victoria Cross awarded to a pilot of RAF Bomber Command was that won by Flight-Lieutenant R A B Learoyd. The award was made for gallantry when, on the night of 12/13 August 1940, Flight-Lieutenant Learoyd

Wellington bomber fitted with a de-gaussing ring; by flying low over the sea such aircraft were able to explode German magnetic mines harmlessly

was flying Hampden P4403, one of a force of five from Nos 49 and 83 Squadrons which dropped delayed action bombs on an aqueduct of the Dortmund–Ems Canal.

The only Victoria Cross ever to be awarded to a member of RAF Fighter Command was that won by Flight-Lieutenant James Brindley Nicolson, RAF, on 16 August 1940. A Flight Commander of No 249 (Hurricane) Squadron, Nicolson was leading a section of three fighters on patrol near Southampton, Hants, when he sighted enemy aircraft ahead. Before he could complete the attack his section was 'bounced' from above and behind by German fighters which shot down one Hurricane and set Nicolson's aircraft ablaze. With flames sweeping up through his cockpit, the British pilot remained at his controls long enough to complete an attack on an enemy aircraft which had flown into his sights, and then bailed out. Meanwhile on the ground a detachment of soldiers, seeing Nicolson and his wingman descending on parachutes and believing them to be enemy paratroops, opened fire with rifles. Nicolson was hit but survived his wounds and burns, but his colleague was dead when he reached the ground (whether or not he was killed by rifle-fire has never been established).

The first regular-serving American pilot to die in action during the Second World War was Pilot Officer William M L Fiske, RAF, who on 17 August 1940 died of wounds suffered in action at Tangmere, England, during the Battle of Britain on 16 August 1940.

ALLIED PILOTS WHO SCORED TEN OR MORE CONFIRMED VICTORIES DURING THE BATTLE OF BRITAIN:

Pilot	Victories	Aircraft	Squadron	Notes
Sgt J František, DFM	17	Hurricanes	(303 Sqdn)	Czech. Top-scoring Czech and Allied pilot
Plt Off E S Lock, DSO, DFC*	16+1 shared	Spitfires	(41 Sqdn)	Top-scoring British pilot
Fg Off B J G Carbury, DFC*	15+1 shared	Spitfires	(603 Sqdn)	Top-scoring New Zealand pilot
Sgt J H Lacey, DFM*	15+1 shared	Hurricanes	(501 Sqdn)	Top-scoring Auxiliary pilot
Plt Off R F T Doe, DSO, DFC*	15	Hurricanes / Spitfires	(238 Sqdn) / (234 Sqdn)	British
Flt-Lt P C Hughes, DFC	14+3 shared	Spitfires	(234 Sqdn)	Top-scoring Australian pilot
Plt Off C F Gray, DSO, DFC**	14+2 shared	Spitfires	(54 Sqdn)	New Zealander
Flt-Lt A A McKellar, DSO, DFC*	14+1 shared	Hurricanes	(605 Sqdn)	British
Fg Off W Urbanowicz, DFC	14	Hurricanes / Hurricanes	(303 Sqdn) / (601 Sqdn)	Top-scoring Polish pilot
Fg Off C R Davis, DFC	11+1 shared	Hurricanes	(601 Sqdn)	Top-scoring South African pilot
Flt-Lt R F Boyd, DSO, DFC*	11+1 shared	Hurricanes	(601 Sqdn)	British
Sgt A McDowall, DSO, AFC, DFM*	11	Spitfires	(602 Sqdn)	British
Fg Off J W Villa, DFC*	10+4 shared	Spitfires / Spitfires	(72 Sqdn) / (92 Sqdn)	British
Fg Off D A P McMullen, DFC**	10+3 shared	Spitfires / Spitfires	(54 Sqdn) / (222 Sqdn)	British
Flt-Lt R S S Tuck, DSO, DFC**	10+1 shared	Spitfires / Hurricanes	(92 Sqdn) / (257 Sqdn)	British
Plt Off H C Upton, DFC	10+1 shared	Spitfires / Spitfires	(43 Sqdn) / (607 Sqdn)	Top-scoring Canadian pilot
Flt Sgt G C Unwin, DSO, DFM*	10	Spitfires	(19 Sqdn)	British

(The ranks shown are those held during the Battle of Britain. The decorations are shown to include *all* gallantry awards won by these pilots before, during and after the Battle.)

Convair B-36 six-engined bomber (USAF)

The largest flying-boat to attain operational status in the Second World War was the German Blohm und Voss Bv 222 Wiking. Designed as a transatlantic civil transport for Lufthansa, the prototype (D-ANTE) first flew on 7 September 1940, but was quickly impressed into war service as a cargo transport. The final Bv 222C version had a wing span of 150 ft 11 in (46 m), gross weight of 108 000 lb (48 990 kg) and was powered by six Junkers Jumo 205C engines.

The first parachute-borne magnetic mines to be dropped in British coastal waters during the Second World War, on the night of 20–1 November 1940, were dropped by Heinkel He 115 twin-engined seaplanes of 3/Kü Fl Gr 906 (*Küstenfliegergruppen*). Two nights later a mine of this type was dropped off Southend and, recovered at low tide, was the first to be de-fused and removed for examination.

The first American-built fighter aircraft in British service to destroy a German aircraft in the Second World War were two Grumman Martlets (USN designation F4F-3) of the Royal Navy. Patrolling over Scapa Flow on 25 December 1940, Martlets of No 804 Squadron, flown by Lieutenant L V Carver, RN, and Sub-Lieutenant Parke, RNVR, intercepted and forced down a Junkers Ju 88. Known in US Navy service as the F4F Wildcat, this was Grumman's first monoplane fighter. Its first operational use in USN service was in the defence of Wake Island.

The largest bomber aircraft ordered by the USAAF during the Second World War was the Convair B-36. Its six engines, driving pusher propellers, were mounted within the deep-section wing which spanned 70·10 m (230 ft). The design was evolved to make it possible for attacks to be launched against European targets from North American bases. Two prototypes were ordered in 1941, but production aircraft entered service too late

Grumman Wildcat V of the Fleet Air Arm

Hurricane Mk I of the Merchant Ship Fighter Unit being catapulted from the CAM ship *Empire Tide* (Imperial War Museum)

to be deployed operationally during the Second World War.

The first British single-seat monoplane fighter to serve on board aircraft carriers of the Royal Navy was the Hawker Sea Hurricane. The type equipped No 880 Squadron in January 1941 and was embarked in HMS *Furious* in July of the same year. They were used also aboard merchant ships and a number of small naval catapult ships under the 'Catafighter' scheme. Their first success in this role came on 3 August 1941 when the Sea Hurricane of HMS *Maplin*, flown by Lieutenant R W H Everett, RNVR, shot down a German Fw 200 Condor.

The first 4000 lb (1814 kg) 'block-buster' bomb to be used on operations was dropped by a Wellington of No 149 Squadron, during an attack on Emden on 1 April 1941.

The first aircraft designed as a jet fighter, and also the first twin-engined jet aircraft, was the German Heinkel He 280. The first prototype, the He 280V-1, first flew on 2 April 1941 powered by two Heinkel HeS 8 turbojets, each developing approximately 1102 lb (500 kg) static thrust. Maximum level speed of the He 280V-5, with HeS 8A engines of 1650 lb (750 kg) static thrust, was demonstrated to be 510 mph (820 km/h) at 19680 ft (6000 m). It did not achieve production status, being abandoned in favour of the Me 262.

The first German turbojet-powered aircraft to enter operational service was the Messerschmitt Me 262A. The first flight of the Me 262V-1 prototype, powered by a single 1200 hp Junkers Jumo piston-engine, was made on 18 April 1941. It was not until 18 July 1942 that the first flight with two turbojet engines was recorded, these being Junkers 109-004A-0 turbojets, each of 1848 lb (840 kg) static thrust. The first production aircraft had 109-004B-1 turbojets of 1980 lb (900 kg) st, which provided a maximum level speed of about 536 mph (868 km/h) at 23 000 ft (7000 m).

The Me 262A-2a Sturmvogel ('Stormbird') fighter-bomber variant is believed to have entered service with Kommando Schenk in early

Grumman TBF Avengers of the US Navy

Messerschmitt Me 262 captured at the end of the war

July 1944, moving to Juvincourt in France on 10 July 1944 to begin operations with six aircraft. This suggests that the Me 262 was **the first turbojet-powered aircraft to enter operational service** in the world. The Me 262A-1a Schwalbe ('Swallow') entered operational service on 3 October 1944; a test unit was expanded and renamed Kommando Nowotny – under the command of the Austrian ace, Major Walter Nowotny – and became operational on that date. One of the two Staffeln of the unit was based at Achmer, the other at Hesepe.

The largest airborne assault mounted by the Luftwaffe during the Second World War was Operation 'Mercury', the landing of 22750 men on the island of Crete commencing at 07.00 h on 20 May 1941. The Luftwaffe used 493 Junkers Ju 52/3m aircraft and about 80 DFS 230 gliders. The assault was made by 10000 parachutists, 750 troops landed by glider, 5000 landed by Ju 52/3ms and 7000 by sea. The operation, although regarded as a brilliant success, cost Germany about 4500 men killed and some 150 transport aircraft destroyed or badly damaged, and effectively brought Luftwaffe paratroop operations to an end.

The first combat mission flown by the Boeing B-17 Flying Fortress was a daylight raid flown at 30000 ft (9150 m) against Wilhelmshaven by three aircraft of No 90 (Bomber) Squadron, RAF on 8 July 1941. Twenty B-17Cs had been supplied to the RAF, and were used by this service under the name Fortress I. From 1942 onwards about 200 B-17F and B-17G aircraft were delivered to the RAF, which designated the 'Fs' as Fortress II and IIA, the 'Gs' as Fortress III. All of the II and IIA aircraft served with Coastal Command's Very Long Range force for mid-Atlantic patrol.

The first aircraft of twin-boom configuration to enter service with the USAAF, in August 1941, was the Lockheed P-38 Lightning. It was also that service's first twin-engined single-seat fighter, and the first squadron fighter aircraft to be equipped with turbochargers.

Regarded as the most successful USAAF fighter aircraft in the Pacific theatre of operations during the Second World War, the Lockheed P-38 Lightning had the distinction also of destroying more Japanese aircraft than any other USAAF fighter in that theatre. Major Richard Bong scored all 40 of his accredited victories while flying P-38s.

The first single-engined American aircraft equipped with a power-operated gun-turret, and also the first to carry a 22 in torpedo, was the Grumman TBF Avenger. First flight of an XTBF-1 prototype was made on 1 August 1941. First operational use of production aircraft was at the Battle of Midway on 4 June 1942, when five out of six aircraft deployed were lost. Despite this inauspicious start, the Avenger became one of the most outstanding naval aircraft of the Second World War. Almost 10000 were built, of which nearly 1000 served in 15 first-line squadrons of the British FAA.

The first RAF fighter capable of exceeding a speed of 400 mph (644 km/h) was the Hawker Typhoon, which entered squadron service with No 56 at Duxford in September 1941. Armed with four 20 mm cannon, and able to carry two 1000 lb bombs or eight 60 lb rocket-projectiles beneath its wings, the Typhoon be-

Lockheed P-38 Lightning, known to the Germans as the "Fork-tail Devil"

Hawker Typhoon Mk IB (Charles E Brown)

De Havilland Mosquito Mk IV bombers of the Royal Air Force (Charles E Brown)

came famous for 'train-busting' activities, and devastated German *Panzer* divisions at Caen and the Falaise gap after the Allied invasion of Europe.

A classic aircraft, and the finest American fighter produced during the Second World War, North American's P–51 Mustang first entered service with the British Royal Air Force in November 1941. Merlin-engined Mustangs were first delivered to the United States 8th Air Force in the UK on 1 December 1943, and flew their first long-range escort mission, to Kiel, on 13 December 1943. Their first mission to Berlin, escorting B–17s and B–24s, was made in March 1944.

The fastest aircraft in RAF Bomber Command for an entire decade, from November 1941 until introduction of the English Electric Canberra in 1951, was the de Havilland Mosquito. Entering squadron service with No 105 at Swanton Morley, Mosquitos made their first operational sortie on 31 May 1942, four aircraft making a surprise attack on Cologne just a few

North American P-51H Mustang fighter

hours after the first 1000-bomber raid. Too fast to be intercepted during much of its wartime service, the Mosquito had the lowest loss rate of any aircraft in Bomber Command. Fighter variants of the Mosquito were no less successful, the Mk VI being the most extensively built, entering service with Fighter Command as a day and night intruder. Mosquitos served also as night fighters, responsible for home defence, and on the night of 14/15 June 1944 **the first VI flying-bomb to be shot down was destroyed** over the English Channel by a Mosquito of No 605 Squadron flown by Flight-Lieutenant J G Musgrave.

The first bombs to fall on American targets in the Second World War were dropped on 7 December 1941 by Aichi D3A Type 96 carrier bombers of the Japanese Navy. A total of 123 D3A1s took part in this first surprise attack, bombs from some of these aircraft falling on US military installations on Oahu Island just before the first bombs began to fall on Pearl Harbor.

The first United States aircraft to be involved in offensive action in the Second World War were B–17 Flying Fortresses which, on 10 December 1941, attacked Japanese shipping. The United States 8th Air Force made its **first attack against the German homeland** on 27 January 1943, when B–17Fs were deployed against Wilhelmshaven. On 4 March 1944 B–17Gs were used for the **USAAF's first attack on Berlin.**

The first German aircraft to be destroyed by the USAAF, within minutes of America's declaration of war against Hitler's Germany on 11 December

1941, was a Focke-Wulf Fw 200 Condor shot down over the North Atlantic. The victory was scored by a P–38 Lightning operated by the Iceland-based 342nd Composite Group.

The first successful operational sortie against the Japanese to be flown by the American Volunteer Group (AVG) – better known as General Claire Chennault's Flying Tigers – and during which six of ten attacking bombers were destroyed, was made on 20 December 1941. Based in China, primarily to defend the Burma Road, the AVG was credited with the destruction of 286 Japanese aircraft from its formation in December 1941 to its absorption into the USAAF on 4 July 1942.

The last of the famous American Curtiss fighters to serve with the RAF was the Kittyhawk, which entered service late in 1941. A total of more than 3000 of these aircraft were delivered to Commonwealth air forces.

The first combat operation carried out by Avro Lancaster heavy bombers was a mine-laying sortie flown by No 44 (Bomber) Squadron, based at Waddington, Lincolnshire, over the Heligoland Bight on 3 March 1942. **Their first night-bombing attack** was recorded when two of No 44 Squadron's aircraft took part in a raid on Essen on the night of 10/11 March 1942. The first of many famous raids involved 12 aircraft of Nos 44 and 97 Squadrons, led by Squadron Leader J D Nettleton, which made a low-level daylight attack on the MAN Diesel factory at Augsburg on 17 April 1942. Their first operation with the Pathfinder Force was made on the night of 18–19 August 1942.

The first naval battle in which the issue was decided by aircraft alone was the Battle of the Coral Sea, fought on 7–9 May 1942, between US Navy Task Force 17 and Vice-Admiral Takeo Takagi's Carrier Striking Force (part of Vice-Admiral Shigeyoshi Inouye's Task Force MO). The battle was fought to prevent Japanese support of an invasion of Port Moresby and disrupt Japanese plans to launch air strikes against the Australian mainland. In this respect the battle must be considered to have been an American victory, although the large American carrier USS *Lexington* was sunk – **the first American carrier to be lost in the Second World War.** The opposing carrier forces were as follows:

US Navy:

USS *Lexington*	23 F4F Wildcat fighters, 36 SBD Dauntless dive-bombers and 12 TBD Devastator torpedo-bombers.
USS *Yorktown*	21 F4F Wildcat fighters, 38 SBD Dauntless dive-bombers and 13 TBD Devastator torpedo-bombers.

Imperial Japanese Navy:

Shoho	12 A6M Zero fighters, 9 B5N Kate torpedo-bombers.
Shokaku	21 A6M Zero fighters, 21 D3A Val dive-bombers and 21 B5N torpedo-bombers.
Zuikaku	21 A6M Zero fighters, 21 D3A Val dive-bombers and 21 B5N torpedo-bombers.

The Japanese carrier *Shoho* was attacked and sunk by Dauntless and Devastator aircraft from the *Lexington* and *Yorktown*, becoming **the first Japanese aircraft carrier to be destroyed by American airmen** and giving rise to the famous radio call from Lieutenant-Commander Robert Dixon 'Scratch one flat-top'. A total of 69 American naval aircraft were lost during the battle while the Japanese losses amounted to about 85 as well as about 400 naval airmen (many of whom went down with the *Shoho*). The Japanese carrier *Shokaku* was also damaged severely, but was able to limp home for repairs. The loss of experienced airmen and the absence of the *Shokaku* critically weakened Japanese naval forces that were to be involved in the vital Battle of Midway, fought between 4 and 7 June 1942.

The most successful US Navy fighter aircraft of the Second World War was the Grumman F6F Hellcat, the prototype of which flew for the first time on 26 June 1942. US Navy statistics record that almost 75 per cent of all wartime combat victories were achieved with Hellcats, of which 12 275 had been built when production ended in November 1945.

The largest and heaviest twin-engined aircraft to enter service with the USAAF, the Curtiss C–46 Commando, was evolved from a 36-passenger commercial transport designed by the Curtiss-Wright company in 1937. More than 3000 of the military transport versions were built for the USAAF, the first entering service in July 1942, and the majority serving in the Pacific theatre. They made a significant contribution to operations in that area after the loss of Burma and the Burma Road to the Japanese, carrying supplies to China over the Himalayan "Hump" route from India.

Curtiss C-46 Commando transport over the "Burma Hump"

The first Boeing B-17E Flying Fortress to arrive in Britain was allocated to the USAAF's 97th Bombardment Group. This unit made its first operational sortie in Europe on 17 August 1942, when 12 B-17Es attacked Rouen.

The highest interception by an unpressurised aircraft, its pilot G W H Reynolds unaided by a pressure suit and breathing only a conventional oxygen supply, was that made at 49 500 ft (15 090 m) in a specially prepared Spitfire VC operating from No 103 MU near Alexandria, in late August 1942. A Ju 86P-2 high-altitude pressurized aircraft was destroyed in this interception.

The first fighter aircraft of US manufacture to be ferried directly across the North Atlantic, from the USA to the UK via Labrador, Greenland and Iceland, were Lockheed P-38 Lightnings allocated for service with the USAAF in Europe. Ferried initially by pilots of the 1st and 14th Fighter Groups in the late summer of 1942, they were provided with B-17 escorts for navigational assistance.

Perhaps the most unusual aircraft to enter Luftwaffe service, in late 1942, was the Heinkel He 111Z. It consisted of two conventional He 111s, linked together by a new section of wing, which carried a fifth engine. Intended to tow the Messerschmitt Me 321 *Gigant* glider, the He 111Z saw very limited service in this role and as a transport aircraft.

The first US-designed and -built jet fighter was the Bell P-59 Airacomet, the prototype of which flew for the first time on 1 October 1942. One of

these aircraft was exchanged with Britain for a Gloster Meteor 1, to allow each nation to evaluate the design of these early turbine-powered aircraft.

The largest and heaviest single-engined single-seat fighter adopted by the USAAF at the time of its entry into service, with the 56th Fighter Group in November 1942, was the Republic P-47 Thunderbolt. Its maximum take-off weight exceeded that of two Supermarine Spitfires. The XP-47 prototype first flew in 1940, and when production ended a total of 15683 had been built. The Thunderbolt was **the last radial-engined fighter to serve in quantity with the USAAF,** and examples were to remain in service with America's Air National Guard squadrons until 1955.

The first four-engined transport aircraft to serve with the USAAF, the Douglas C-54 Skymaster, was evolved from the civil DC-4 designed for the long-range routes of US commercial operators. Entering service in December 1942, they achieved a remarkable record, establishing regular and reliable services across the North Atlantic to the UK, across the Indian Ocean between Australia and Ceylon, and from the US west coast over the Pacific to the Philippines and Australia. From the time they entered service until VJ-Day, they recorded 79642 long-range ocean crossings, and during this period only three aircraft were lost.

Probably the most amazing airborne interception of the Second World War was that achieved by pilots of the USAAF's 12th, 70th and 339th Squadrons on 18 April 1943. At 0935 hrs a group of P-38G Lightnings intercepted on schedule two Mitsubishi G4M1 bombers (*Bettys*) over Bougainville – escorted by six Mitsubishi A6M

America's first jet fighter, the Bell Airacomet (foreground), with a piston-engined Bell P-63 Kingcobra

Dornier's 'push-and-pull' fighter, the Do 335 Pfeil

Zeros (*Zekes*) – which were carrying Admiral Isoroku Yamamoto and members of his staff on an inspection itinerary. Both *Bettys* were destroyed and Admiral Yamamoto, who had master-minded the attack on Pearl Harbor, was killed according to plan.

The best known, and probably the most widely used transport aircraft ever to serve with the USAAF, the Douglas C-47 Skytrain was a militarised version of the Douglas DC-3 civil transport which had revolutionised US domestic routes in the late '30s. Entering USAAF service in 1941, the Skytrain (usually known affectionately as the "Gooney Bird") served all over the world in almost every theatre of war. C-47s played a conspicuous role in operations involving airborne troops, dropping 4381 paratroops over Sicily on 10 June 1943, during the first large-scale Allied airborne operation. Skytrains were also involved heavily in the return to Burma and the D-Day assault on Normandy.

The first jet bomber to enter operational service with any air force – apart from the Me 262A-2a (see above), which was merely a fighter with two bomb pylons fitted beneath the nose – was the Arado Ar 234B Blitz ('Lightning'), powered by two Jumo 004B series turbojets. The first prototype Ar 234V1 made its first flight on 15 June 1943, powered by two Jumo 004A turbojets of 1850 lb (839 kg) thrust. First to enter service were small numbers of Ar 234B-Os, issued to various Luftwaffe reconnaissance units during the summer of 1944. The first operational use of the Ar 234B in the bombing role was during the Ardennes offensive, which began on 16 December 1944 and was considered to be over on 16 January 1945. The Blitz was a significant aircraft, carrying up to 4400 lb (2000 kg) of bombs and with a maximum speed of

461 mph (742 km/h) at 19685 ft (6000 m). It entered service too late to have any influence on the air war in Europe.

The most successful Russian woman fighter pilot of the Second World War, and thus presumably the most successful woman fighter pilot in the world, served with the mixed-sex 73rd Guards Fighter Air Regiment. She was Junior Lieutenant Lydia Litvak, and she was killed in action on 1 August 1943 at the age of 22 with a total of 12 confirmed victories to her name. She flew Yak fighters.

(During the Second World War, most Russian women combat pilots served with the 122nd Air Group of the Soviet Air Force. This all-female unit comprised the 586th Fighter Air Regiment, the 587th Bomber Air Regiment and the 588th Night Bomber Air Regiment. The 586th IAP (Istebitelnyi aviatsionnyi polk =fighter air regiment) was formed at Engels, on the Volga River, in October 1941; it was commanded by Major Tamara Aleksandrovna Kazarinova. The pilots of this unit flew a total of 4419 operational sorties, took part in 125 air combats, and were credited with 38 confirmed victories. The unit flew Yak-1, -7B and -9 fighters. During the Second World War, thirty Russian airwomen received the gold star of a Hero of the Soviet Union. It is believed that 22 of them served with the 588th/ 46th Guards Night Bomber Air Regiment, which was equipped with Po-2 biplanes.)

The fastest piston-engined fighter designed for the Luftwaffe, the Dornier Do 335, demonstrated a maximum speed of 474 mph (763 km/h) at 21 000 ft (6400 m). First flown in September 1943, the Do 335 Pfeil had a unique engine layout, one Daimler-Benz DB 603 being mounted conventionally in the nose, with a second DB603 mounted in the rear fuselage and driving a pusher propeller through an extension shaft. It was developed too late to enter operational service.

The first specially-designed anti-submarine patrol aircraft for the Japanese Navy was the Kyushu Q1W Tokai, the prototype of which made its first flight in September 1943. Only a few Q1W1s entered service before the war's end, equipped with radar and magnetic anomaly detection (MAD) equipment.

The first major operational success by a guided free-falling (i.e. unpowered) bomb was the sinking of the Italian battleship *Roma* by Dornier Do 217s of 111 Gruppe, Kampf-

Large formations of American B-17 Fortresses of the 390th Bomb Group, 13th Bomber Wing, 8th Air Force, appear to make tempting targets; but vapour trails high above indicate the presence of a strong escort of P-47 Thunderbolts

geschwader 100, commanded by Major Bernhard Jope, west of Corsica on 9 September 1943, using Ruhrstahl Fritz-X 3100 lb (1406 kg) bombs. The *Roma* was hit by two bombs, the second of which started a disastrous fire which reached the magazine and caused the battleship to blow up, break in two and sink with most of her crew. In this Italian fleet, which was *en route* to surrender to the Allies at Malta, the *Roma*'s sister ship, the *Italia*, received a direct hit on the bows and took on about 800 tons (813 tonnes) of water before reaching Malta under her own steam. Fritz-X bombs later scored hits on the battleship HMS *Warspite*, the British cruisers HMS *Uganda* and *Spartan* (sunk), and the American cruiser USS *Savanna*.

The first single-seat fighters in the world to enter service with air interception (AI) radar were Vought F4U-2 Corsairs of the United States Navy. Twelve aircraft formed the initial night fighter units in 1943, serving with Navy Squadrons VFN-75 and VFN-101. On 31 October 1943, one of these aircraft operating over New Georgia was credited with the Navy's first interception achieved solely by airborne radar.

The first operational use of Grumman F6F-3E Hellcat night fighters occurred on 26 November 1943, when two aircraft from the USS *Enterprise* dispersed Japanese bomber formations attacking a US task force during a night landing on the Gilbert Islands.

The first operational carrier-based fighter in the world with a tricycle undercarriage was the Grumman F7F-1 Tigercat. The first of two XF7F-1 prototypes made its first flight in December 1943, but rapidly changing requirements and alterations to specification meant that

US Marine Corps Vought Corsair on Bougainville (US Navy)

production aircraft entered squadrons too late to see operational service in the Second World War. A two-seat night-fighter variant (the F7F-2N) was also produced.

The largest-calibre multi-barrelled weapon fired by an aircraft was the six-barrelled 77 mm Sondergeräte 113A Forstersonde rocket mortar fitted in three Henschel Hs 129 ground-attack aircraft to fire Sabot-type shells vertically downwards. The weapon was triggered by a photo-electric cell actuated by the shadow of a tank beneath the aircraft.

The last scout aeroplane of the US Navy was the Curtiss SC Seahawk, designed for operation from battleships, carriers and from land bases. First flown on 16 February 1944, more than 500 were built for the Navy before cancellation of outstanding contracts after VJ-day. The SC-1s were supplied in a landplane configuration and were converted to floatplanes, as required, in Navy workshops.

Top scorers in the Battle of the Philippine Sea, 19-20 June 1944, were the US Navy's F6F Hellcats which were credited with the destruction of most of the 300-plus aircraft lost by the Japanese in this action. Other achievements in this category made the Hellcat the **most significant Allied shipboard fighter of the Second World War,** credited with the destruction of more than 5000 enemy aircraft. This massive total was achieved for the loss of fewer than 300 Hellcats, to give a kill/loss ratio of more than 19 to 1.

The most extensively built version of the North American Mustang was the P-51D, which entered production in mid-1944, with almost 8000 examples completed. It featured an all-round-vision bubble canopy on a cut-down rear fuselage, was armed with six 0·5 in machine-guns mounted in the wings, and was powered by a 1695 hp Packard-built V-1650-7 Merlin engine.

The first operational military use of composite combat aeroplanes was by the German Luftwaffe, during the Allied liberation of France in June–July 1944. Devised originally under a programme designated 'Beethoven-Gerät', it was known subsequently as the 'Mistel-Programm'. Biggest problem was to develop an effective system by which the pilot of the single-seater upper aircraft could control and effect separation of the two components. Initial operational Mistel 1s comprised an upper piloted Bf 109F-4 and a lower Ju 88A-4 which carried a warhead containing 3800 lb (1725 kg) of high explosive. The weapon was first issued to 2 Staffel of Kampfgeschwader 101, commanded by Hauptmann Horst Rudat, which was formed in April 1944. The first operational use of the device, known unofficially as Vater und Sohn (Father and Son), was on the night of 24/25 June 1944, when five Mistels were deployed against Allied shipping in the Seine Bay. Later versions of the composite had Fw 190s as the upper component.

The only known pilot who has been both jailed and awarded his country's highest gallantry decoration for the same exploit was Lieutenant Michael Devyatayev, a Soviet fighter pilot shot down by the Luftwaffe over Lwow on 13 July 1944. Taken prisoner by the Germans, Devyatayev escaped, seized a Heinkel He 111 bomber and flew nine other escapees back to Russian-held territory. On regaining his freedom the 23-year-old pilot was gaoled under the USSR criminal code which labelled him a traitor for having been taken prisoner. Nine years later, in 1953, he was freed under an amnesty prevailing at the time, and in 1958 was made Hero of the Soviet Union and awarded the Order of Lenin and Gold Star Medal.

The first (and only swept-wing) rocket-engined aeroplane to enter operational squadron service with any air force was the Messerschmitt Me 163B-1 Komet interceptor fighter, powered by a Walter 109-509A-2 rocket motor, using the liquid propellants known as T-Stoff (hydrogen peroxide and water) and C-Stoff (hydrazine hydrate, methyl alcohol and water) to give a maximum static thrust of 3300 lb (1500 kg). Maximum speed of the Komet in combat was about 600 mph (965 km/h). The swept-wing, tailless Me 163B-1a was armed with two 30 mm MK 108 cannon; some machines are known to have carried various experimental armament systems in addition. The Komet equipped only one combat unit, Jagdgeschwader 400, which comprised eventually three Staffels. The whole unit was concentrated on Brandis in July 1944, and the first operation was flown against a group of B-17s on 16 August 1944, without any of the US bombers being destroyed. Although approximately

300 of these rocket-powered interceptors were built, JG400 claimed only nine Allied aircraft destroyed and two probables before the units were disbanded in early 1945.

The first loss of a jet aircraft in aerial combat is thought to have taken place on 28 August 1944, when Major Joseph Myers and Lieutenant M Croy of the 78th Fighter Group, US 8th Air Force, were credited with the destruction of an Me 262 operated by the Kommando Schenk.

The shortest elapsed time for the development of an entirely new jet fighter (which achieved combat status) was 69 days for the Heinkel He 162 Salamander. Conceived in an RLM specification issued to the German aircraft industry on 8 September 1944, the He 162 was the subject of a contract issued on 29 September 1944 for an aircraft capable of being mass-produced by semi-skilled labour using non-strategic materials. Sixty-nine days later, on 6 December 1944, the first prototype He 162V-I was flown by Heinkel's Chief Test Pilot, Kapitän Peter, at Vienna-Schwechat. On 10 December the prototype broke up in the air and crashed before a large gathering of officials, and Peter was killed. Notwithstanding this set-back, the aircraft entered production and joined I and II Gruppen of Jagdgeschwader I at Leck/Holstein during April 1945. III Gruppe of this Geschwader was under orders to receive the new fighter but was forestalled by the end of the war. Known also as the Volksjäger, or 'People's Fighter', it was intended that large numbers would be constructed, but only 116 A-series machines were completed. The Salamander was not a pleasant machine to fly, and as a result few of these aircraft were encountered in combat. Its single BMW 003 turbojet, rated at 1760 lb (800 kg) thrust, provided a maximum level speed of 522 mph (840 km/h) at 19685 ft (6000 m).

The first aviation unit specifically formed for suicide operations was the *Shimpu* Special Attack Corps, a group of 24 volunteer pilots commanded by Lieutenant Yukio Seki, formed within the 201st (Fighter) Air Group, Imperial Japanese Navy, during the third week of October 1944. The unit, equipped with Mitsubishi A6M Zero-Sen single-seat fighters, was formed for the task of diving into the flight decks of American aircraft carriers in the Philippines area, with 550 lb (250 kg) bombs beneath the fuselages of the fighters. (*Shimpu* is an alternative pronun-

Junkers Ju 88P-2, used mainly for anti-tank operations on the Russian front, armed with two 37 mm guns in a belly pack

ciation of the Japanese ideographs which also represent *kamikaze*, 'Divine Wind', the name more generally applied to Japanese suicide operations.) **The first successful suicide attack was carried out** on 25 October 1944 when five Zeros, flown by members of the Special Attack Corps, sank the US escort carrier *St Lo* and damaged the carriers *Kalinin Bay*, *Kitkun Bay* and *White Plains*.

The first aerial victory against another piloted aircraft gained by the pilot of a jet aircraft has never been positively identified, but was certainly achieved in the first week of October 1944 by a pilot of Kommando Nowotny, the target being a Boeing B-17 Flying Fortress of the US 8th Air Force.

The first jet-fighter ace in the world has not been positively identified, but it is thought that he was one of the pilots of Kommando Nowotny. The unit was withdrawn from operations following the death in action of Major Walter Nowotny on 8 November 1944, and later provided the nucleus for the new fighter wing, Jagdgeschwader 7 'Nowotny'; III Gruppe, JG 7 became operational during December 1944. Hauptmann Franz Schall is known to have scored three aerial victories on the day of Nowotny's death, and subsequently served with 10 Staffel, JG 7; it is therefore entirely possible that he was the first pilot in the world to have achieved five confirmed aerial victories while flying jet aircraft. Other known jet aces of the Second World War are listed below. The fragmentary records which survived the final immolation of the Luftwaffe in 1945 prevent the preparation of a complete list, and the following should therefore be regarded simply as a framework for future research:

Oberstleutnant Heinz Bär (JV 44)   .   .   16
Hauptmann Franz Schall (10/JG 7)   .   .   14
Major Erich Rudorffer (II/JG 7) .   .   .   12
Oberfeldwebel Hermann Buchner (III/JG 7) .   12
Leutnant Karl Schnörrer (II/JG 7)   Not fewer than 8
Leutnant Rudolf Rademacher
(II/JG 7)   .   .   .   .   .   Not fewer than 8
Major Theodor Weissenberger (Staff/JG 7) .   8
Oberleutnant Walter Schuck (3/JG 7) .   .   8
Oberst Johannes Steinhoff (Staff/JG 7, JV 44)   6
Major Wolfgang Späte (Staff/JG 7) .   .   5
Leutnant Klaus Neumann (JV 44)   .   .   5

Douglas AD-1 Skyraider

**The most successful destroyer of flying
bombs (V-1s) in flight** was Squadron Leader
Joseph Berry, DFC**, who shot down 60 during 1944.

**The largest flying-boat to serve with the
US Navy** was the Martin JRM Mars. Twenty
JRM-1 transports were ordered in January 1945
but only five were built, plus one heavier JRM-
2. With a wing span of 200 ft (60·96 m) and
gross weight of 165 000 lb (74 842 kg) in the
JRM-2, the Mars flying-boat had a maximum
speed of 225 mph (362 km/h). It was converted
after the war for water-bombing of forest fires.

**The largest military flying-boat built in
Germany during the Second World War**
was the Blohm und Voss Bv 238. The single
prototype, powered by six Daimler-Benz DB
603 engines, first flew in March 1945. Intended
to fill a long-range reconnaissance or transport
role, the Bv 238 had a wing span of 197 ft 4½ in
(60·17 m) and a maximum loaded weight of
176 370 lb (80 000 kg).

**The first test-drop of the 22 000 lb (9980 kg)
'Grand Slam' bomb** was made from an Avro

Avro Lancaster Mk II, one of the variants
with Hercules radial engines

Lancaster on 13 March 1945. **The first operational drop of this bomb** was made by Squadron Leader C C Calder of No 617 (Bomber)
Squadron, flying Lancaster B1 (Special) *PD112*.
The bomb was dropped on the Bielefeld Via-
duct on 14 March 1945, smashing two of its
spans.

**The first single-seat carrier-based dive-
bomber and torpedo-carrier in US Navy
service,** the Douglas AD-1 Skyraider, was
developed too late to see operational service in
the Second World War. First flown on 18 March
1945, the Skyraider was to prove an important
naval aircraft in the Korean and Vietnam con-
flicts. When supplied to the British Royal Navy
under the MAP, it filled a unique position as an
airborne early-warning aircraft. Designated
AEW1 in British service, the Skyraiders were
also the last piston-engined fixed-wing aircraft
in first-line service with the FAA.

**The first successful use of a purpose-built
suicide aircraft** is thought to have taken place
on 1 April 1945, when three Yokosuka MXY7
Ohka ('Cherry Blossom') piloted rocket-
powered 'flying bombs' of the 721st Air Group,
Imperial Japanese Navy, were released over an
American naval force near Okinawa (approximate position 26° 15' N, 127° 43' E). Damage
was inflicted on the battleship USS *West Virginia*, the attack cargo-ships *Achernar* and *Tyrell*,
and the attack transport *Alpine*, **The first ship
to be sunk by one of these suicide aircraft**
was the US destroyer *Mannert L Abele*, on 12
April 1945, near Okinawa (approximate position 27° 25' N, 126° 59' E). The first operational
deployment of these suicide aircraft, on 21
March 1945, had been a complete failure. The
16 Mitsubishi G4M2e launch aircraft were intercepted short of their target and forced to jettison
their piloted weapons.

V1 flying bomb photographed by the camera-gun of an attacking RAF fighter

The first land-based fighter strikes against Tokyo, carried out by North American P-51Ds of the United States 7th Air Force operating from Iwo Jima, were made on 7 April 1945. Some 80 Mustangs escorted 300-plus Superfortresses in attacks on aircraft factories at Tokyo and Nagoya.

The last U-boat to be sunk by RAF Coastal Command aircraft (the 196th) was destroyed by a Consolidated Catalina of No 210 Squadron on 7 May 1945. It was an RAF Catalina of No 209 Squadron which, on 26 May 1941, spotted the German battleship *Bismarck* after surface vessels had lost contact.

The only combat aircraft of canard configuration to be the subject of a production contract in the Second World War was the Japanese Kyushu J7W Shinden, intended as a heavily armed high-performance interceptor for use by the Navy. Powered by a 2130 hp Mitsubishi MK9D 18-cylinder supercharged radial engine, driving a six-blade pusher propeller, the prototype made its first flight on 3 August 1945. Only two more short flights were made before the Japanese surrender.

The only Japanese turbojet-powered aircraft to take off under its own power during the Second World War was the Nakajima Kikka. Designated Navy Special Attacker Kikka, its design was based on the Messerschmitt Me 262, but its development came so late in the war that the prototype flew only twice, on 7 and 11 August 1945.

The last sortie by suicide aircraft, according to Japanese accounts, was flown on 15 August 1945 by seven aircraft of the Oita Detachment, 701st Air Group, Imperial Japanese Navy, led in person by Admiral Matome Ugaki, commander of the 5th Air Fleet. United States records fail to confirm any *kamikaze* attacks on this date however.

'Little Boy', the 10 ft long, 9000 lb atomic bomb detonated over Hiroshima, 6 August 1945

Col. Paul W Tibbets (centre), pilot of the B-29 Superfortress *Enola Gay* from which the first operational atomic bomb was dropped over Hiroshima

'Fat Man', 10 ft 8 in long and weighing 10 000 lb, was the second atomic bomb, detonated over Nagasaki, Japan, on 9 August 1945

Smoke billows up over Nagasaki after the atomic attack by just two B-29s, one the bomb-carrier, the other an escort

The total number of suicide aircraft expended, and the results of these attacks, are believed to be as follows:

	Sorties	Aircraft returned	Expended
Philippines area	421	43	378
Formosa area	27	14	13
Okinawa area	1809	879	930
Total	2257	936	1321

It has not proved possible to distinguish between actual suicide aircraft and escort fighters in the Okinawa operations and this must necessarily invalidate the total figures to some extent. A rough estimate would show that the usual ratio of escort fighters to suicide aircraft on most sorties was about three to two, although late in the campaign many sorties were flown entirely without escort.

The total number of American naval vessels sunk by suicide attacks from the air was 34, and 288 damaged. Those which were sunk comprised 3 escort aircraft carriers, 13 destroyers, 1 destroyer escort, 2 high-speed minelayers, 1 submarine chaser, 1 minesweeper, 5 tank-landing ships, 1 ocean tug, 1 auxiliary vessel, 1 patrol craft, 2 motor torpedo-boats and 3 other vessels.

The most successful fighter pilot in the world, and Germany's leading ace in the Second World War, was Major Erich Hartmann of Jagdgeschwader 52. Born in Weissach, Württemberg, on 19 April 1922, he was still a schoolboy when war broke out. It was 10 October 1942 before he was posted to his first combat unit, 9 Staffel of JG 52, which was operating in the Ukraine. This unit had earned the reputation of being one of the most formidable Staffeln in the Luftwaffe; in the spring of 1942 its pilots, led by the ace Oberleutnant Hermann Graf, had been credited with 47 victories in 17 days. In this rarefied atmosphere Hartmann, just 20 years old, did not give any immediate signs of future promise. He gained his first victory on 5 November, but by April 1943, when he had amassed 100 missions in his log-book, his score stood at only seven victories. Like many other leading aces, he had a lengthy 'running-in' period during which he perfected his technique. Nor did he strive for high scores on each sortie, preferring to gain one good, clean textbook kill and then concentrate on his flying and his rear-view mirror. Nevertheless, he was to gain multiple victories on many occasions, his first major success being achieved on 7 July 1943; on that day Hartmann led 7 Staffel's Messerschmitt Bf 109G fighters from their base at Ugrim, to score seven personal victories in three sorties – three Ilyushin Il-2 ground-attack aircraft, and four Lavochkin LaGG-3 fighters. His Knight's Cross came with his 148th victory on 29 October 1943, the Oak Leaves on 2 March 1944 with his 200th, and the Swords on 4 July 1944, with his 239th. The

Ilyushin Il-2s – the famous Shturmovik tank-busters of the Soviet air force, described by Stalin as being as essential to the Red Army as air and bread

summer of 1944 saw another period of multiple successes; in four weeks he destroyed 78 enemy aircraft, including eight on 23 August, and 11 on the following day, bringing his score to 301 – and making him the first of the only two fighter pilots in the world who ever scored 300 victories. This feat brought him into the select band of men – numbering 27 only – who wore the Diamonds to the Knight's Cross, the award being made on 25 July 1944. In October 1944 he became Staffelkapitän of 4/JG 52, and took over command of II Gruppe, JG 52 on 1 February 1945. His unit retreated steadily westwards as the Red Army swept across central Europe in the final great offensive, and Hartmann eventually surrendered to American forces in Czechoslovakia in May 1945; by this time he had scored 352 victories.

The most successful German fighter pilot in combat against the Western Allies during the Second World War was Hauptmann Hans-Joachim Marseille, who was born in Berlin-Charlottenburg in December 1919. In April 1941 he was posted to I Gruppe Jagdgeschwader 27 in Libya, and it was in desert warfare that he excelled. He worked with great perseverance to master his trade, and in the blinding skies of North Africa his superb vision and marksmanship became a legend. He was credited with many multiple victories, including the astounding total of 17 aircraft destroyed on 1 September 1942. He was awarded the Knight's Cross on 22 February 1942 for his 50th victory; the Oak Leaves on 6 June, for his 75th; and the Swords only 12 days later, by which time his score stood at 101. On 2 September 1942 he received the Diamonds, then his country's highest award, and was only the fourth man to receive it. On 30 September the 'Star of Africa' died; he was forced to bale out when the engine of his Messerschmitt Bf 109G began to smoke for no apparent reason. He is thought to have been struck by the tail of his aircraft as he jumped, and his parachute was not seen to open. He was 22 years old, and had been credited with 158 victories, all of them gained in combat against the RAF and Commonwealth air forces.

The most successful English fighter pilot of the Second World War was Group Captain James Edgar 'Johnnie' Johnson, credited with 38 confirmed aerial victories over German aircraft: Johnson applied to join the Royal Air Force in 1937, but was rejected. In 1939, when the need for aircrew was receiving priority, he was invited to apply once more, and two days later was a Flight Sergeant in the Royal Air Force Volunteer Reserve. Towards the end of August 1940 he was posted to No 19 Squadron, then based at Duxford, the famous fighter station near Cambridge. There was no time to train tyro pilots at Duxford, and Johnson was transferred to No 616 Squadron, then going through a rest and reorganisation period at Coltishall after being withdrawn from combat. He scored his first solo victory in June 1941 – a Messerschmitt Bf 109, shot down over Gravelines. During this period Johnson was flying in the leading section of the three-squadron Tangmere Wing, under the leadership and tutelage of the legless Douglas Bader. His score rose steadily but not spectacularly, and he was awarded the DFC and given command of No 610 Squadron in July 1942. It was while leading this mixed-nationality unit that he began to emerge as one of the brightest stars of Fighter Command. An enforced period of non-combat duty followed a period in command of the Kenley Wing, and it was March 1944 before Johnson returned to operational flying; he was given command of No 144 Canadian Wing, and in his last air battle – a sortie over Arnhem on 27 September 1944 – he scored his 38th kill. Johnson remained in the Royal Air Force, retiring finally with the rank of Air Vice-Marshal in 1966; during the Korean War he secured an exchange posting to the USAAF and flew several combat missions. His decorations include the DSO and two Bars, the DFC and Bar, the American DFC, Air Medal and Legion of Merit, the CB and the CBE. From an international point of view, Johnson certainly destroyed the **greatest number of German fighters** of any Allied pilot.

Messerschmitt Bf 109E

One that did not get away – a B-17 of the 483rd Bomb Group is hit by anti-aircraft fire

The most successful American fighter pilot of the Second World War was Major Richard Ira Bong, whose 40 confirmed aerial victories are unsurpassed by any American military pilot of any war. Born at Superior, Wisconsin, on 24 September 1920, Bong enlisted as a Flying Cadet on 29 May 1941. After flying training at Tulare and Gardner Fields, Calif, and Luke Field, Ariz, he received his 'wings' and a commission (all American military pilots were automatically commissioned) on 9 January 1942. In May he was posted to Hamilton Field, Calif, for combat training on the Lockheed P-38 Lightning twin-engined fighter, and subsequently joined the 9th Fighter Squadron of the 49th Fighter Group, then based in Australia. All of his 40 victories had been scored by late 1944, in the Pacific theatre of war, and General George C Kenney, his commanding officer, ordered him back to the United States in December 1944, with a recommendation for the Congressional Medal of Honor – which award was subsequently granted. Bong became a test pilot for Lockheed at Burbank, Calif; and on 6 August 1945, the

day the world's first atomic bomb was dropped on Hiroshima, he died when the engine of his P-80 jet failed. Many of his victories were gained while flying the P-38J *Marge* named after his fiancée, which is illustrated on page 145.

The numbers of aircraft shot down by fighter pilots of the Second World War varied much more widely than was the case in the First World War, due to the enormous differences in conditions and standards of equipment in the various combat areas. Comparison of the lists of national top-scoring fighter pilots reveals the almost incredible superiority of German pilots in terms of confirmed victories – i.e. Major Erich Hartmann, the Luftwaffe's leading ace, is credited with nearly nine times as many victories as the leading British and American pilots, and 35 Germans are credited with scores in excess of 150.

Since the end of the war there have been persistent attempts to discredit these scores; but by any reasonable criterion, the figures must now be accepted as accurate. The Luftwaffe's confirmation procedure was just as rigorous as that followed by Allied air forces, and the quoted figures are those prepared at unit level and were not subject to manipulation by the Propaganda Ministry. The main reasons for the gulf between German and Allied scores were the different conditions of service and the special circumstances which existed on the Russian Front in 1941 and 1942. In Allied air forces an operational tour by a fighter pilot was almost invariably followed by a posting to a second-line establishment for several months. This process of rotating pilots to areas where they could recover from the strain of prolonged combat operations was unknown in the Luftwaffe; apart from very short periods of leave, a German fighter pilot was effectively on combat operations from the day of his first posting until the day his career ended – in death, serious injury or capture. The Luftwaffe fighter pilot's career was thus, in real terms, about twice as long as his RAF or USAAF counterpart.

When Germany invaded the Soviet Union in June 1941, the Russian Air Forces were equipped with very large numbers of obsolescent aircraft. They had no fighter whose speed and armament approached the performance of the Messerschmitt Bf 109E and Bf 109F, and their bombers in squadron service were markedly inferior to contemporary European designs. Thus, the Luftwaffe was presented with large numbers of easy targets – the perfect environment for the development of a fighter pilot's skill and confidence. The situation did not become signifi-

cantly more challenging for many months, by which time many of the Jagdflieger had learned their trade so well that they retained the initiative. Despite this factor, one is left with the inescapable conclusion that Germany simply produced a group of officers who were fighter pilots of exceptional skill and determination.

The pilots who scored *100 or more victories against the Western Allies* in northern Europe, southern Europe, the Mediterranean area and North Africa were as follows (Western victories only, in cases of mixed service):

Hauptmann Hans-Joachim Marseille	.	.	158
Oberstleutnant Heinz Bär .	.	.	124
Oberstleutnant Kurt Bühligen	.	.	112
Generalleutnant Adolf Galland .	.	.	104
Major Joachim Müncheberg	.	.	102
Oberstleutnant Egon Mayer .	.	.	102
Major Werner Schroer	.	.	102
Oberst Josef Priller .	.	.	101

These figures become even more impressive if one reflects on the fact that Marseille achieved 151 of his victories between April 1941 and September 1942; and that Galland did virtually no combat flying between November 1941 and the end of 1944, while he occupied the post of General of Fighters.

Two categories of victories in northern Europe are worthy of special attention; those scored over heavy bombers, and those scored while flying jet aircraft. The achievements of the world's first generation of jet combat pilots are described elsewhere in this chapter. The Luftwaffe placed great value on the destruction of the

Martin JRM-2 Mars

very heavily armed four-engined Boeing Fortress and Consolidated Liberator bombers which formed the United States 8th Air Force's main equipment in the massive daylight bombing offensive of 1943–5. Usually flying in dense formations protected by an enormous combined firepower – and, in the later months, by superb escort fighters – these large aircraft were obviously far more difficult to destroy than smaller aircraft. The leading 'heavy bomber specialists' among Germany's daylight home defence pilots included:

Oberleutnant Herbert Rollwage . . .	44
Oberst Walther Dahl	36
Major Werner Schroer	26
Hauptmann Hugo Frey	26
Oberstleutnant Egon Mayer . . .	25
Oberstleutnant Kurt Bühligen . . .	24
Oberstleutnant Heinz Bär	21
Hauptmann Hans-Heinrich König . .	20
Hauptmann Heinz Knoke	19

The most successful fighter pilots of the Second World War, by nationality, are listed below: all scores are levelled down to the nearest unit: British gallantry decorations are quoted:

Country of origin		Aircraft destroyed in combat
Australia	Gp Capt Clive R Caldwell, DSO, DFC* . .	28
Austria	Maj Walter Nowotny .	258
Belgium	Flt-Lt Vicki Ortmans, DFC	11
Canada	Sqdn Ldr George F Buerling, DSO, DFC, DFM* .	31
Czechoslovakia	Sgt Josef František, DFM .	28
Denmark	Gp Capt Kaj Birksted either 8 or 10	
Finland	F/Mstr E I Juutualainen .	94
France	Sqdn Ldr Pierre H Clostermann, DFC* .	19
Germany	Maj Erich Hartmann .	352
Hungary	2nd Lt Dezjö Szentgyörgyi	43
Ireland	Wg Cdr Brendan E Finucane, DSO, DFC**	32
Italy	Maj Adriano Visconti .	26
Japan	Sub-Officer Hiroyoshi Nishizawa . .	103
Netherlands	Lt-Col van Arkel 12 V-1s, and 5	
New Zealand	Wg Cdr Colin F Gray, DSO, DFC** .	27
Norway	Flt-Lt Svein Heglund either 14 or 16	
Poland	Jan Poniatowski (rank unknown) . .	36

Romania	Capt Prince Constantine Cantacuzino .	60
South Africa	Sqdn Ldr M T St J Pattle, DFC* . .	41
United Kingdom	Gp Capt James E Johnson, DSO**, DFC* .	38
United States	Maj Richard I Bong .	40
USSR	Guards Col Ivan N Kozhedub . .	62

Fighter pilots serving with the Royal Air Force during the Second World War who achieved 25 or more confirmed aerial victories (countries of origin indicated in parentheses):

Sqdn Ldr M T St J Pattle, DFC*	41	(SA)
Gp Capt J E Johnson, DSO**, DFC*	38	(UK)
Gp Capt A G Malan, DSO*, DFC*	35	(SA)
Wg Cdr B E Finucane, DSO, DFC**	32	(Ir)
Sqdn Ldr G F Buerling, DSO, DFC, DFM* . . .	31	(Ca)
Wg Cdr J R D Braham, DSO**, DFC**, AFC . .	29	(UK)
Wg Cdr R R S Tuck, DSO, DFC** .	29	(UK)
Sqdn Ldr N F Duke, DSO, DFC**, AFC . . .	28	(UK)
Gp Capt C R Caldwell, DSO, DFC*	28	(Au)
Gp Capt F H R Carey, DFC**, AFC, DFM . . .	28	(UK)
Sqdn Ldr J H Lacey, DFM* . .	28	(UK)
Wg Cdr C F Gray, DSO, DFC** .	27	(NZ)
Flt-Lt E S Lock, DSO, DFC* .	26	(UK)
Wg Cdr L C Wade, DSO, DFC**	25	(US)

Fighter pilots serving with the United States air forces during the Second World War who achieved 25 or more confirmed aerial victories:

USAAF:

Major Richard I Bong (CMH) . .	40
Major T B McGuire (CMH) . .	38
Colonel F S Gabreski . . .	31
Lieutenant-Colonel R S Johnson. .	28
Colonel C H MacDonald . .	27
Major G E Preddy . . .	26

USN:

Captain D McCampbell . . .	34

USMC:

Major J J Foss	26
Lieutenant R M Hanson . . .	25
Lieutenant-Colonel G Boyington .	22

(Lieutenant-Colonel Boyington is known to have destroyed an additional six enemy aircraft while serving with the Air Volunteer Group under Chinese command.)

LUFTWAFFE FIGHTER PILOTS WITH 150 OR MORE CONFIRMED VICTORIES DURING THE SECOND WORLD WAR AND THE SPANISH CIVIL WAR

E = Eastern Front; W = Europe; Afr = North Africa; Gr = Greece; * = at least

✠🌿⚔◆ = Knight's Cross with Oak Leaves, Swords and Diamonds
✠🌿⚔ = Knight's Cross with Oak Leaves and Swords
✠🌿 = Knight's Cross with Oak Leaves
✠ = Knight's Cross of the Iron Cross

Name, rank, decorations	Units	Total score	Day/ Night	Fronts	Four engined	With jet a/c
Major Erich Hartmann ✠🌿⚔◆	JG 52	**352**	352/0	352 E	o	o
Major Gerhard Barkhorn ✠🌿⚔	JG 52, 6, JV 44	**302**	301/1	301 E	o	?
Major Günther Rall ✠🌿⚔	JG 52, 11, 300	**275**	275/0	3 W, 272 E	?	o
Oberleutnant Otto Kittel ✠🌿⚔	JG 54	***267**	267/0	267 E	o	o
Major Walter Nowotny ✠🌿⚔◆	JG 54, Kdo. Nowotny	**258**	258/0	255 E, 3 W	*1	3
Major Wilhelm Batz ✠🌿⚔	JG 52	**237**	237/0	232 E, 5 W	2	o
Major Erich Rudorffer ✠🌿⚔	JG 2, 54, 7	**222**	222/0	136 E, 60 W, 26 Afr	10	12
Oberstleutnant Heinz Bär ✠🌿⚔	JG 51, 77, 1, 3, JV 44	**220**	220/0	96 E, 79 W, 45 Afr	*21	16
Oberst Hermann Graf ✠🌿⚔◆	JG 51, 52, 50, 11, 52	**212**	212/0	202 E, 10 W	10	o
Major Theodor Weissenberger ✠🌿	JG 77, 5, 7	**208**	208/0	175 E, 33 W	?	8
Oberstleutnant Hans Philipp ✠🌿⚔	JG 76, 54, 1	**206**	206/0	177 E, 29 W	1	o
Oberleutnant Walter Schuck ✠🌿	JG 5, 7	**206**	206/0	198 E, 8 W	4	8
Major Heinrich Ehrler ✠🌿	JG 5, 7	***204**	204/0	204 E?	?	?
Oberleutnant Anton Hafner ✠🌿	JG 51	**204**	204/0	184 E, 20 Afr	5	o
Hauptmann Helmut Lipfert ✠🌿	JG 52, 53	**203**	203/0	majority E, *4 W	2	o
Major Walter Krupinski ✠🌿	JG 52, 5, 11, 26 JV 44	**197**	197/0	177 E, 20 W	1	?
Major Anton Hackl ✠🌿⚔	JG 77, 11, 26, 300, 11	**192**	192/0	105 E, 87 W	32	o
Hauptmann Joachim Brendel ✠🌿	JG 51	**189**	189/0	189 E	o	o
Hauptmann Max Stotz ✠🌿	JG 54	**189**	189/0	173 E, 16 W	*o	o
Hauptmann Joachim Kirschner ✠🌿	JG 3, 27	**188**	188/0	167 E, 13 Gr, 6 W, 2 Malta	*2	o
Major Kurt Brändle ✠🌿	JG 53, 3	**180**	180/0	160 E, 20 W	o	o
Oberleutnant Günther Josten ✠🌿	JG 51	**178**	178/0	majority E	1	o
Oberst Johannes Steinhoff ✠🌿⚔	JG 26, 52, 77, 7, JV 44	**176**	176/0	148 E, 28 W & Afr	4	6
Oberleutnant Ernst-Wilhelm Reinert ✠🌿⚔	JG 77, 27	**174**	174/0	103 E, 51 Afr, 20 W	2	o
Hauptmann Günther Schack ✠🌿	JG 51, 3	**174**	174/0	174 E	o	o
Hauptmann Emil Lang ✠🌿	JG 54, 26	**173**	173/0	148 E, 25 W	?	o
Hauptmann Heinz Schmidt ✠🌿	JG 52	**173**	173/0	173 E	o	o
Major Horst Ademeit ✠🌿	JG 54	**166**	166/0	165 E, 1 W	o	o
Oberst Wolf-Dietrich Wilcke ✠🌿⚔	JG 53, 3	**162**	162/0	137 E, 21 W, 4 Malta	4	o
Hauptmann Hans-Joachim Marseille ✠🌿⚔◆	JG 52, 27	**158**	158/0	151 Afr, 7 W	o	o
Hauptmann Heinrich Sturm ✠	JG 52	**157**	157/0	157 E	o	o
Oberleutnant Gerhard Thyben ✠🌿	JG 3, 54	**157**	157/0	152 E, 5 W	?	o
Oberleutnant Hans Beisswenger ✠🌿	JG 54	**152**	152/0	152 E	o	o
Leutnant Peter Düttmann ✠	JG 52	**150**	150/0	150 E	o	o
Oberst Gordon Gollob ✠🌿⚔◆	ZG 76, JG 3, 77	**150**	150/0	144 E, 6 W	o	o

Republic P-47D Thunderbolt, 42-26418, *flown from Boxted, England, during the summer of 1944 by Lieutenant-Colonel Francis S Gabreski, commanding officer of the 61st Fighter Squadron, 56th Fighter Group, United States 8th Air Force. Gabreski's final score of thirty-one confirmed aerial victories qualifies him as America's leading ace in the European theatre of operations.*

Grumman F6F-5 Hellcat 'Minsi III', *flown from the carrier USS Essex during the summer of 1944 by Commander David McCampbell of Fighter Squadron VF-15. McCampbell's final score of thirty-four confirmed aerial victories qualifies him as the US Navy's leading ace of the Second World War.*

Lockheed P-38J Lightning, 42-103993 'Marge', *flown between October 1943 and March 1944 by Captain Richard I Bong, at that time an Assistant Operations Officer at the headquarters of the US 5th Fighter Command in New Guinea. Bong's final score of forty confirmed aerial victories qualifies him as America's leading ace of the Second World War.*

Nakajima B5N2 'Kate' *in which Commander Mitsuo Fuchida, General Commander (Air) of the Imperial Japanese Navy 1st Carrier Division, led the first wave of the attack on Pearl Harbor on 7 December 1941, and from the cockpit of which he transmitted the order to attack at 07.49 h that morning.*

Mitsubishi A6M2 Reisen (Zero-Sen) *flown in July 1942 from Lae, New Guinea by Petty Officer First Class Saburo Sakai of the Tainan Kokutai, Imperial Japanese Navy Air Force. Sakai, who scored sixty confirmed aerial victories in China and the Pacific before being seriously wounded over Guadalcanal in August 1942, finished the war as Japan's third ranking, and senior surviving, fighter pilot.*

Aircraft flown by outstanding airmen of the Second World War

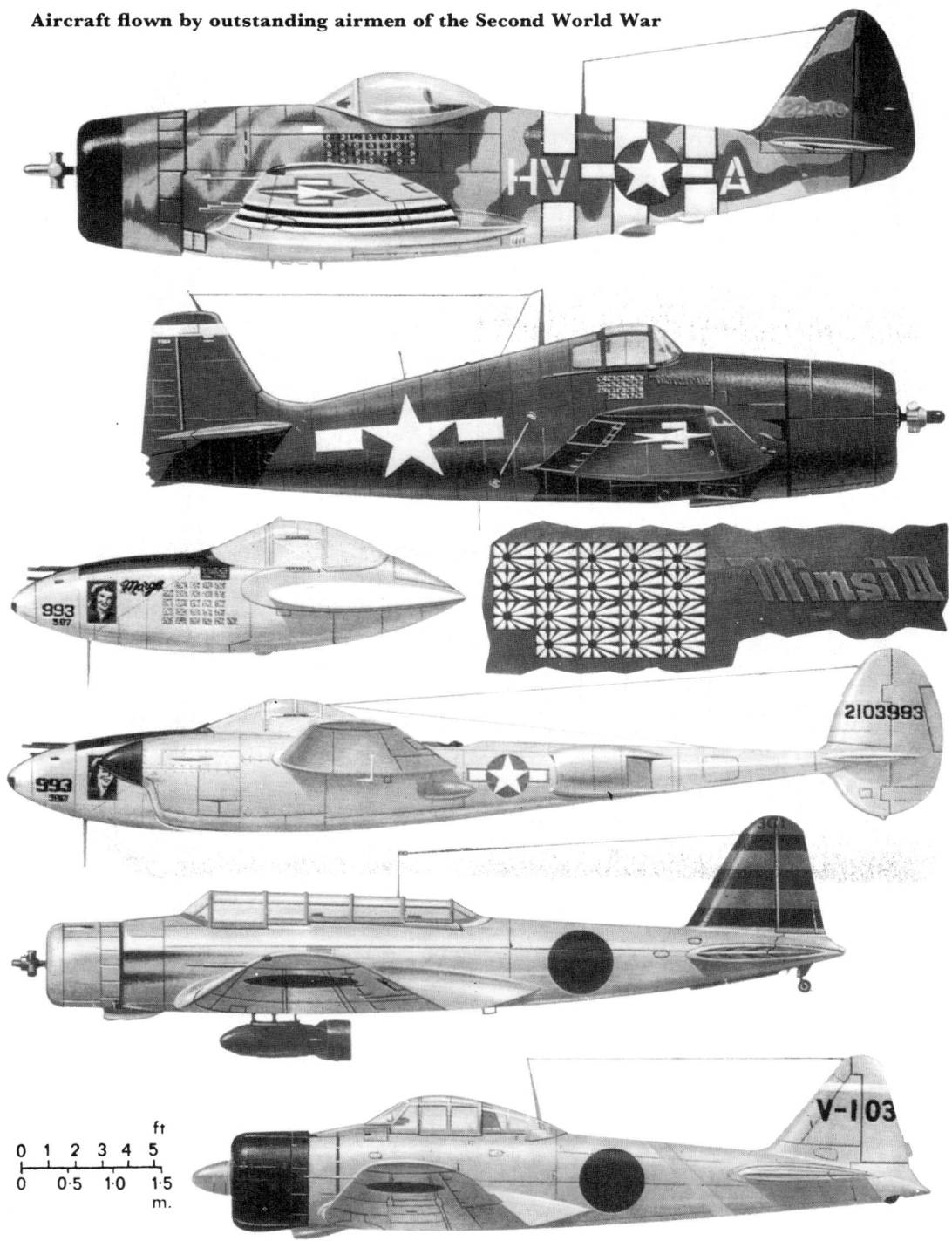

Hitler's 'People's Fighter', the Heinkel He 162 (Royal Aircraft Establishment)

Japanese Kamikaze pilot about to crash his Zero fighter into the USS *Missouri*, April 1945

Concorde, the first supersonic transport to carry passengers in scheduled service

SECTION 8

POST WAR - CIVIL

In retrospect it is fascinating to remember how proud the world's nations were of the civil transport services they could offer in 1939: yet little more than three million passengers were carried by air in that year, and the seating capacity of the average civil airliner was only about sixteen.

Europe, locked in war from September 1939, was concerned with the development of short to medium range fighter and bomber aircraft. America, which came into head-on collision with Japan after that nation's surprise attack on Pearl Harbor on 7 December 1941, was faced with fighting a war on the other side of the Pacific Ocean. Of necessity the US had to develop long-range bomber and cargo aircraft, and this meant that when peace returned again to the world in 1945, the American aircraft industry was in a very strong position to supply the world's airlines with the new civil transport aircraft which they would need.

Britain's lead in the development of gas turbine engines meant that this nation's aircraft industry was the first to provide the world with turboprop- and turbojet-powered airliners. It was but a short-term lead, for the failure of the Comet 1 enabled America to draw level again and introduce into service the superb Boeing 707. From it has stemmed the whole family of Boeing aircraft, up to the present-day Model 747, the world's first 'jumbo-jet' or, as properly termed, wide-body airliner. Alongside the Boeing development have come similar aircraft from the Douglas Company and, more recently, from Lockheed.

In the field of supersonic transport (SST) aircraft the Anglo-French Concorde has an undoubted lead over the rest of the world, simultaneous inaugural flights by Air France and British Airways on 21 January 1976 marking the beginning of supersonic intercontinental travel for fare-paying passengers. Russia's Tu-144 SST, though first to fly as a prototype on 31 December 1968, is not expected to begin passenger-carrying services until some time in 1978.

The expansion of air travel provided by the big commercial airlines has been mirrored by a similar growth in business and commuter aircraft, they too benefitting from new technology and improved power plants. But the enormous growth in private flying has been slowed by the world shortage of hydrocarbon fuels and their rapidly escalating cost.

Nonetheless, the demand for air services continues to grow, and the prime task of the world's aviation industries will be to improve operating economies by one means or another. With a steadily increasing workload falling on air traffic control we may well see the introduction of even bigger 'jumbos' in the not too distant future.

The first post-war British survey flight to South America was made on 9 October 1945, when Captain O P Jones took off from Hurn in the Lancastrian G-AGMG.

The first regular British air service to South America was inaugurated on 15 March 1946, initially using Lancastrians.

The first transatlantic arrivals at London's Heathrow Airport, opened officially on 31 May 1946, were Lockheed Constellations of Pan American Airways and American Overseas Airlines.

British European Airways (BEA) began its first scheduled all-cargo services on 10 August 1947.

The first helicopter-operated public mail service in the United Kingdom was inaugurated on 1 June 1948, flown by BEA Sikorsky S-51s.

Sikorsky S-51 of British European Airways, operated on the pioneer helicopter mail services in 1948

The first car ferry flight operated by Silver City Airways was made on 14 July 1948. Bristol Freighter G-AGVC made the initial flight, carrying two cars.

First cross-Channel vehicle ferry service by Silver City Airways, 14 July 1948

Prototype Vickers Viscount, first turbine-powered airliner to carry passengers in scheduled service

Saunders-Roe Princess flying-boat

The United Kingdom's first night airmail services to be operated by helicopter began on 17 October 1949, the inaugural flight being made by BEA's Sikorsky S-51 G-AJOV, flown by Captain J Cameron.

The first sustained and regular scheduled helicopter passenger services in the United Kingdom, between Liverpool and Cardiff, began on 1 June 1950.

The first turbine-powered airliner in the world to receive an Airworthiness Certificate was the Vickers V630 Viscount prototype, with Rolls-Royce Dart turboprop engines, which was awarded Certificate No A907 on 28 July 1950. The following day British European Airways operated this aircraft to record **the world's first scheduled passenger service to be flown by a gas-turbine-powered airliner.** Piloted by Captain Richard Rymer, the Viscount (*G-AHRF*) took off from London (Northolt) and flew to Paris (Le Bourget) carrying 14 fare-paying passengers and 12 guests of the airline. Captain Rymer was also **the world's first holder of a pilot's licence for a turbine-powered civil transport aircraft.**

The first freight service operated by turbo-prop-powered aircraft, between Northolt

and Hanover, was flown by Rolls-Royce Dart-engined Douglas DC-3s of BEA. The first flight was made on 15 August 1951 by G-ALXN *Sir Henry Royce*. This aircraft and a second DC-3, G-AMDB *Claude Johnson*, had been used for development flying of the Dart power plants introduced on the Vickers Viscount.

The world's first turbojet airliner to enter airline service was the de Havilland DH106 Comet 1 powered by four de Havilland Ghost 50 turbojet engines. **The world's first regular passenger service to be flown by turbojet aircraft** was inaugurated by British Overseas Airways Corporation on 2 May 1952 using the de Havilland Comet 1 *G-ALYP* between London and Johannesburg, South Africa. Its route was via Rome, Beirut, Khartoum, Entebbe and Livingstone, and the aircraft was captained in turn by Captains A M Majendie, J T A Marsden and R C Alabaster. It carried 36 passengers and the total elapsed time for the 6724 miles (10 821 km) was 23 h 34 min.

The first, and the only one to fly, of the three giant Saunders-Roe SR45 Princess flying-boats, made its maiden flight on 20 August 1952, piloted by Geoffrey Tyson. Because of the shift from flying-boats to landplanes for long-range intercontinental air services, the three Princesses were scrapped.

The world's first fatal accident involving a turbojet airliner occurred on 3 March 1953 when *Empress of Hawaii*, a de Havilland Comet 1 *CF-CUN*, of Canadian Pacific Air Lines crashed on take-off at Karachi, Pakistan, during its delivery flight from London to Sydney, its intended operating base. All 11 occupants were killed, and it was stated that the accident was caused by the pilot lifting the nose too high during take-off, thereby causing the aircraft to stall.

de Havilland Comet 1 airliner operated by Air France

The first post-war landing in the United Kingdom of a German civil-registered aircraft was made on 15 April 1955. The aircraft was the Deutsche Lufthansa Convair CV-340 *D-ACAD* on a route proving flight. The first regular post-war air services to the UK operated by Lufthansa began on 16 May 1955.

The first non-stop flight by an airliner from London to Vancouver, on the Pacific coast of Canada – a distance of 5100 miles (8208 km) – was completed by the turboprop-powered Bristol Britannia 310 *G-AOVA* on 29 June 1957.

The first Transatlantic passenger service to be flown by turbine-powered airliners was inaugurated by BOAC on 19 December 1957 using Bristol Britannia 312 turboprop aircraft. The first flight from London to New York was by *G-AOVC* captained by Captain A Meagher.

The first Transatlantic service by turbojet airliners was operated by BOAC with de Havilland Comet 4s on 4 October 1958. Simultaneous flights were made in each direction, Captain R E Millichap being responsible for the London–New York service in *G-APDC*, and Captain T B Stoney for the New York–London flight with *G-APDB* completed in a record time of 6 h 11 min.

The last scheduled flight by an Airspeed Ambassador of BEA's Elizabethan class

First swing-tail freighter, the Canadair CL-44D-4

was made on 30 July 1958. In six years the 20-strong fleet had flown a combined distance of about 31 million miles (50 million km) and had carried almost 2½ million passengers.

The only transport aircraft in the world with a sales total exceeding 1000 examples is the Boeing Model 727 three-turbofan short/medium range airliner. Initial design began in June 1959, and by 1 February 1977 a total of 1364 had been sold, of which 1242 had been delivered.

The first round-the-world passenger service by jet airliners was established by Pan American World Airways during October 1959. The first aircraft flown on this service was a Boeing 707-321, *Clipper Windward*.

Boeing 727, first jet airliner to exceed 1000 sales, over Mount Rainier, Washington State, USA

Tupolev Tu-144, first supersonic airliner to fly (Tass)

The world's first aircraft to be fitted with a hinged tail for rear loading, the Canadair CL-44D-4, was flown for the first time on 16 November 1960. Powered by four Rolls-Royce Tyne 515/10 two-shaft two-spool turbo-props, each rated at 5730 ehp, it had a range of 5660 miles (9110 km) with a 37300 lb (16920 kg) payload. Its maximum payload was 66048 lb (29960 kg).

The world's first supersonic transport aircraft to fly was the Soviet Union's Tupolev Tu-144, the prototype of which flew for the first time on 31 December 1968. On 26 May 1970 this prototype became the **world's first commercial transport aircraft to exceed a speed of Mach 2** by flying at 1335 mph (2150 km/h) at a height of 53475 ft (16300 m). The maiden flight of the prototype was also **the first time that its Kuznetsov NK-144 turbofan engines had been tested in the air.**. Then rated at 28660 lb (13000 kg) static thrust without afterburning and 38580 lb (17500 kg) st with afterburning, the four turbofan engines of production aircraft are each rated at 44090 lb (20000 kg) st with afterburning. **Regular supersonic flights, the first in the world,** were begun by Aeroflot on 26 December 1975. These were, however, confined to the carriage of freight and mail, and passenger services had not started by early 1977.

The first wide-body commercial transport aircraft in the world was the Boeing 747 'jumbo-jet', the first of which flew for the first time on 9 February 1969 from Paine Field, near Seattle, Washington. **The first commercial service with the 747** was inaugurated by Pan American Airways on its New York/London route on 22 January 1970. Between that date and 1 February 1977, Boeing 747s in airline service around the world had flown more than 4.16 million hours, carried more than 134 million passengers and had flown more than 2172 million miles (3496 million km). By the same date a total of 316 of these aircraft had been ordered, of which 296 had been delivered. Basic seating capacity is for 385 passengers, but a ten-abreast high-density arrangement can accommodate 500 passengers. A long-range lighter-weight version, the Boeing 747SP, with accommodation for 305/321 passengers, flew for the first time on 4 July 1975. When the first of these aircraft for South African Airways was delivered, on 23–4 March 1976, its delivery flight from Paine Field, Washington, to Cape Town, a distance of 10290 miles (16560 km), represented a world record for non-stop distance flown by a commercial aircraft.

The world's first supersonic commercial transport aircraft to operate regular scheduled passenger services is the British Aircraft Corporation/Aérospatiale Concorde. The Aérospatiale 001 prototype made its first flight on 2 March 1969, and the BAC 002 flew for the first time on 9 April 1969. **The Concorde test**

programme was the most comprehensive ever undertaken for a civil airliner, involving eight flight aircraft plus two airframes for structural ground testing. Just prior to the type's entry into service, test aircraft had flown more than 5500 h, of which more than 2000 h were at supersonic speed. In route proving, test and demonstration flights, Concordes had then landed at 83 airports in 49 countries and had flown more than 5 million miles (8 million km).

The first passenger services were flown on 21 January 1976 when simultaneous take-offs were made by Air France's 205 from Paris to Rio de Janeiro, via Dakar, and British Airways' 206 from London to Bahrain.

The first wide-body commercial transport aircraft produced by the aircraft industry of Europe is the Airbus Industrie A300 Airbus, the first of which flew for the first time on 28 October 1972. Bought initially by Air France, it entered service with this airline on its Paris–London route on 23 May 1974. A truly international project, the A300 is built by Aérospatiale of France, Deutsche Airbus (MBB and VFW-Fokker) of Germany, Fokker-VFW of the Netherlands, CASA of Spain and Hawker Siddeley Aviation in the United Kingdom. Powered by two General Electric CF6-50C turbofan engines, each of 51 000 lb (23 130 kg) st, this short/medium-range airliner has a maximum capacity of 345 passengers in a nine-abreast high-density seating arrangement.

CRIME IN THE SKY

Disasters in the Air

The two worst known disasters resulting from criminal violence in aircraft while in flight both involved the *loss of 44 lives.*

On 1 November 1955 a Douglas DC-6B, operated by United Air Lines, exploded in the air and crashed near Longmont, Colorado, USA, killing all 44 occupants. It was later established that John G Graham had introduced a bomb aboard in an insurance plot to murder his mother who was a passenger.

On 7 May 1964 a Fairchild F-27, operated by

Pacific Airlines, crashed near Doublin, Calif, USA, and all 44 occupants were killed. A tape recording indicated that the pilot was shot by an intruder in the airliner's cockpit.

The first known use of aircraft for violence in civil crime was the dropping of three small bombs on 12 November 1926 by an aeroplane on a farmhouse in Williamson County, Illinois; the raid was carried out by a member of the Shelton gang against members of the rival Birger gang in a Prohibition feud involving illicit supply of beer and rum. The bombs however failed to detonate.

Mass hijacking. A BOAC VC10, a TWA Boeing 707, and a Swissair DC-8 were blown up by Palestinian guerrillas at Dawson's Field, a desert airstrip in Jordan, on 12 September 1970 after being hijacked. Total value of the three aircraft was about £8 500 000 (United Press International)

Lightning Mk 2A of No 19 Squadron, RAF Germany

SECTION 9

POST WAR - MILITARY

No comparison of the rate and scope of development of military aircraft during the two World Wars is really possible. This is due primarily to the fact that the geographical scale of the two wars was vastly different. The First World War, in the main, was confined to air activity in Europe and western Asia: it fostered the development of combat aircraft and medium range bombers.

The Second World War, fought over a worldwide arena, made essential the development of fighter and bomber aircraft with greater range capability, and of cargo aircraft which could carry men and their weapons and supplies over vast distances. In the process came the evolution of more powerful engines, of improved structures, the development of radar navigation equipment, better communications equipment and the first indications of far more potent weapons.

The achievement of these improvements was not accidental. Some had come from the inspiration imposed by the stress of war: the majority stemmed from painstaking research which, once initiated, left a bank of basic research from which post-war engineers would be able to borrow and develop new ideas and techniques. Certainly the most exciting invention to come to fruition during the war was the gas turbine engine. Even the most naive and non-technical observer could guess that within a short time new power plants would bring startling changes to military aircraft.

One stream of research had been concerned with the aerodynamic problems that resulted as aircraft approached the speed of sound. Work carried out in Germany had been fairly advanced, and at the war's end this information, when integrated with that done by Allied aviation researchers and engineers, paved the way to startling new performance. In Britain, as a single illustrative example, the 616 mile/h (991 km/h) world speed record set by Group Captain E M Donaldson in a Gloster Meteor was, in the short space of ten years, almost doubled when, on 10 March 1956, Peter Twiss flew the Fairey Delta 2 at a speed of 1132 mile/h (1822 km/h) to establish the first 'over 1000 mile/h' (1600 km/h) world speed record.

Similar high-speed development took place in other fields, particularly in radar and weapons. Space research, stemming from the ballistic rockets which Germany had launched against European and British targets in the latter stages of the war, emphasised the development of electronic equipment. The requirements in this field brought also an urgent need to carry miniaturisation of such equipment to the most advanced state.

All of this has meant that, throughout the post-war period, military aircraft of all nations have reflected the results of this research in their ever-increasing standards of sophistication. There has, of course, been an ever-increasing price tag on the finished product, and the electronics of some modern aircraft can cost more than the airframe which carries them. In some cases the pilot has little more to do than monitor equipment which flies the aircraft to its target, releases its weapons, and then flies it back to its base. Neither he, nor the taxpayers who pay for such devastating weapons, will complain if their kill potential is so horrifying that it is sufficient to prevent any new major war.

The fastest twin piston-engined combat aircraft in the world to reach operational status was the de Havilland Hornet fighter which possessed a maximum speed of 485 mph (780 km/h) in 'clean' combat configuration. Powered by two 2070 hp Rolls-Royce Merlin 130 engines, the Hornet was armed with four 20 mm guns, could carry up to 2000 lb (907 kg) of bombs or rockets on under-wing pylons, and had a maximum range of over 2500 miles (4025 km). It was first flown by Geoffrey de Havilland, Jr, on 28 July 1944, but did not reach the first RAF squadron – No 64 (Fighter) Squadron at Horsham St Faith, Norfolk – until May 1946,

after the end of hostilities in Europe. The Hornet was **the fastest piston-engined fighter to serve with the RAF** and also the **last piston-engined fighter to serve with RAF first-line squadrons.**

The last piston-engined fighter in FAA first-line squadrons was the Hawker Sea Fury. First flown on 21 February 1945, the type entered service with No 807 Squadron in August 1947 and operated with distinction throughout the Korean War. The Sea Fury flown by Lieutenant P Carmichael of No 802 Squadron destroyed the squadron's first MiG-15 on 9 August 1952.

C-47 transports of the USAF being unloaded at Tempelhof Airport during the Berlin Airlift of 1948–49 – the first occasion on which war was averted by the large-scale use of unarmed aircraft

McDonnell FH-1 Phantoms, first jets to be based on an aircraft carrier

de Havilland Vampire, piloted by Lieutenant Commander E M Brown, taking off from HMS *Ocean* 3 December 1945 (Charles E Brown)

The first twin-engined single-seat fighter to operate from aircraft carriers of the Royal Navy was the de Havilland Sea Hornet F20. Aircraft of this type equipped No 801 Squadron, and were embarked in HMS *Implacable* in 1949, remaining in service until the squadron was re-equipped with Sea Furies in 1951. The prototype Sea Hornet (*PX212*) made its first flight on 19 April 1945. Production aircraft were powered by two 2030 hp Rolls-Royce Merlin 133s or 134s, and had a maximum speed of 467 mile/h (742 km/h) at 22000 ft (6700 m).

The first American aeroplane to land under jet power on a ship was a Ryan FR-1 Fireball compound fighter, powered by a conventionally mounted Wright R-1820-72W radial piston-engine, as well as by a General Electric I-16 turbojet installed in the rear fuselage. This combination had resulted from the US Navy's doubts of the suitability of jet-powered aircraft for carrier operations. Flown by Ensign Jake C West on to the escort carrier USS *Wake Island* on 6 November 1945, it had been intended to fly on using the reciprocating engine, but this failed on the approach and West landed under jet power. **The first US (all-)jet aircraft to land on an aircraft carrier,** on 21 July 1946, was the McDonnell FD-1 Phantom prototype, which landed on the USS *Franklin D Roosevelt*. Production aircraft were subsequently re-designated FH-1.

The first two post-war world absolute air speed records were established by Gloster Meteor F4 fighters. On 7 November 1945 Group Captain H J Wilson established a record speed of 606 mph (975 km/h) at Herne Bay, Kent, flying the Meteor EE454 *Britannia*. On

7 September 1946 Group Captain E M Donaldson raised the record to 616 mph (991 km/h) near Tangmere, west Sussex in Meteor EE549. The three Meteors allocated to the re-formed RAF High Speed Flight (EE548–50), which was to make the attempt to set a new speed record in 1946, had their wings clipped, reducing wing span from 43 ft 0 in (13·11 m) to 37 ft 2 in (11·33 m). Unfortunately, it was discovered that this reduced their maximum speed by almost 58 mph (93 km/h) and full-span wings were used for the record attempt. Because the clipped-wing modification improved structural integrity, as well as rate of roll, it subsequently became standard on all but the earliest F4s. In 1948 Meteor F4s superseded F3s in the RAF's first-line fighter squadrons until they, in turn, were supplanted by F8s. In May 1950, the F4s which equipped No 222 Squadron at Leuchars became the first jet fighters to be based in Scotland. Meteor F8s first entered service with No 245 Squadron at Horsham St Faith, Norfolk, on 29 June 1950. The Meteor F4 was powered by two 3500 lb (1587 kg) thrust Rolls-Royce Derwent 5 engines, the F8 by 3600 lb (1633 kg) thrust Derwent 8s. Meteor F8s of the Royal Australian Air Force were the only British jet fighters to see action in the Korean War.

The world's first pure jet aircraft to operate from an aircraft carrier was a de Havilland Vampire I, the third prototype of which (*LZ551*) had been modified for deck-landing trials. It was first landed on HMS *Ocean*, a light fleet carrier of the Colossus Class, by Lieutenant-Commander E M Brown, RNVR, on 3 December 1945. The first deck landing was followed by trials in

The Yakovlev Yak-15

which 15 take-offs and landings were made in two days.

The first jet fighter to enter squadron service with the Soviet Air Forces was the Yak-15, designed by Alexander S Yakovlev. It entered service with the IA-PVO early in 1947, powered by a single RD-10 turbojet (a Russian adaptation of the German Jumo 004B engine) developing initially 1875 lb (850 kg) thrust. Armed with two 23 mm Nudelman-Suranov NS-23 guns, the Yak-15 had a top speed of 488 mph (786 km/h) at 16400 ft (5000 m). Like the first Tupolev jet bombers, the Yak-15 was also the result of adapting a piston-engine airframe for jet propulsion. The prototype retained the wings, cockpit, tailplane and tailwheel landing gear of a Yakovlev Yak-3, the new engine being mounted in the forward fuselage. This meant that the jet efflux was below the pilot's cockpit, and production aircraft had the fuselage under-surface protected by heat-resistant stainless steel. It meant also that the first batch of production aircraft had an all-metal tailwheel. This proved unsatisfactory, and the Yak-15 was retrofitted with a tricycle-type landing gear. First flight of the Yak-15, with test-pilot M I Ivanhov at the controls, was made on 24 April 1946. A member of the company test team in 1947 was Olga

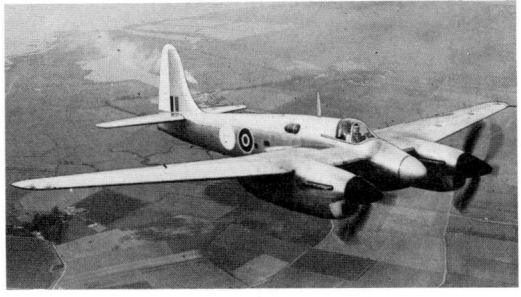

Short Sturgeon (Flight International)

Yamschikova, probably the first woman in history to fly a turbojet-powered fighter aircraft.

The first high-speed twin-engined strike aircraft designed specifically to operate from aircraft carriers of the Royal Navy was the Short Sturgeon. Originating from a wartime requirement, to operate from *Ark Royal* and *Hermes* class carriers, the war's end resulted in the Sturgeon being completed to satisfy a gunnery training and target-towing role. The first prototype made its first flight on 7 June 1946.

The first American pure jet aeroplane to operate from a carrier was the McDonnell FH-1 Phantom, the first jet fighter to serve with first-line squadrons of the US Navy and the US Marine Corps.

The last biplane in squadron service with the Fleet Air Arm was the Supermarine Sea Otter, a carrier-based or shore-based amphibian, of which production ended in July 1946. Used primarily for air–sea rescue and communications, its retirement was hastened by the introduction of helicopters to fulfil these roles.

The first post-war world long distance record for aeroplanes was set up by a modified Lockheed P2V-1 Neptune maritime reconnaissance aircraft, the *Truculent Turtle*, which flew a distance of 11236 miles (18082 km) in September 1946. On 7 March 1949 a later-version P2V-2 took off from the carrier USS *Coral Sea* at the then record take-off weight from a ship of 74000 lb (33566 kg).

The first Russian jet bomber to achieve limited production status was the Tupolev Tu-12, little more than a Tu-2 piston-engined bomber re-engined with gas-turbines. The prototype probably flew for the first time, on 27 July 1947, powered by two 3525 lb (1600 kg) thrust RD-10 engines, derived from the Junkers Jumo 004B. Power plant of the production aircraft consisted of RD-500 engines, Russian equivalent of the Rolls-Royce Derwent 5, which each developed 4410 lb (2000 kg) thrust.

The first flying-boat in the world capable of a maximum level speed of over 500 mph (805 km/h) was the Saunders-Roe SR A/1 jet fighter flying-boat of Great Britain which first flew on 16 July 1947. Powered by two Metrovick Beryl axial-flow turbojets, the SR A/1 had a top speed of 512 mph (824 km/h) and an armament of four 20 mm guns. Three prototypes

Saunders-Roe SR A/1, the first jet fighter flying-boat

were built, but the project was abandoned when flight tests made it clear that the large flying-boat hull compromised both speed and manoeuvrability.

The largest flying-boat in the world, the largest aeroplane ever flown, and the aircraft with the greatest wing span ever built was the Hughes H4 *Hercules*. The 180 ton (183 tonne) flying-boat was powered by eight 3000 hp Pratt and Whitney R-4360 piston-engines, had a wing span of 320 ft (97·54 m), and an overall length of 219 ft (66·75 m). Big enough to accommodate up to 700 passengers, it was intended primarily as a freighter and no cabin windows were provided. Piloted by its sponsor, the American millionaire Howard Hughes, it flew on only one occasion, covering a distance of about a mile, at a height of approximately 80 ft (24 m), over Los Angeles Harbor, Calif, on 2 November 1947.

The first European swept-wing jet fighter to enter operational service after the Second World War was the Swedish Saab J-29 (first flight, 1 September 1948) which joined the Day Fighter Wing F13 of the Flygvapnet near Norrkoping in May 1951. Nicknamed *Tunnan* (Barrel) and powered by a British de Havilland Ghost turbojet of 4410 lb (2000 kg) thrust, the J-29B possessed a top speed of 658 mph (1059 km/h) at 5000 ft (1525 m) or Mach 0·90 at the tropopause.

The first aircraft in the world powered by a coupled twin-turbine engine driving contra-rotating co-axial four-blade propellers was the Fairey Gannet, first flown on 19 September 1949. The unusual power plant was an Armstrong Siddeley Double Mamba turbine, each of its two sections driving one propeller. Half of the engine could be shut down and its propeller feathered to provide a most economical cruise power setting.

The Royal Navy's first all-helicopter squadron was No 705, formed at Gosport in 1950. Equipped with the Westland Dragonfly, the British-built version of the Sikorsky S-51, this type of aircraft quickly demonstrated its value for 'plane-guard' duties and ship-to-shore communications.

First American-built aircraft to enter RAF service after the Second World War were ex-USAF B-29 and B-29A Superfortresses, given the RAF designation Washington B1. No 149 Squadron at Marham was the first squadron to receive these aircraft, in March 1950, and a total of 88 entered RAF service.

The first US Navy jet fighter to take part in air combat was the Grumman F9F-2 Panther, several of which took off from the carrier USS *Valley Forge* off Korea on 3 July 1950 and went into action against North Korean forces. A US Navy pilot of a Grumman Panther shot down a MiG-15 on 9 November 1950 and thus became **the first US Navy jet pilot to shoot down another jet aircraft.** The Panther was the first jet fighter designed by the Grumman Corporation, and the first two XF9F-2 prototypes were powered by imported Rolls-Royce Nene turbojets of 5000 lb (2268 kg) thrust.

The first aerial victory to be gained by the pilot of one jet aircraft over another was achieved on 8 November 1950, when Lieutenant Russell J Brown, Jr, of the 51st Fighter-Interceptor Wing, USAF, flying a Lockheed F-80C, shot down a MiG-15 jet fighter of the Chinese People's Republic Air Force over Sinuiju on the Yalu River, the border between North Korea and China.

Fairey Gannet AS Mk 1 anti-submarine aircraft, with underbelly search radar extended

North American F-86 Sabre of the US Far East Air Forces taking off in Korea. The stripes around the fuselage helped to distinguish F-86s from somewhat similar MiG-15s during hectic air combat

The first turbojet-powered night fighter to enter service with the RAF, serving with No 29 Squadron at Tangmere, Sussex in January 1951, was the Meteor NF11, developed by Armstrong Whitworth.

The first jet aircraft to fly the Atlantic non-stop and unrefuelled was an English Electric Canberra B Mk 2 on 21 February 1951, which was flown from Britain to Baltimore and was later purchased by the USAF to become the first Canberra to carry American markings. Canberras were the first jet bombers produced in Britain and the first to serve with the RAF. The type had the unique distinction of being the first aircraft of non-US design to enter operational service with the USAF after the end of the Second World War. The first of a pre-production batch of eight B-57As (the original USAF designation) made its first flight at Baltimore, Maryland, on 20 July 1953.

The first British V-bomber (so-called from the wing leading-edge plan-form) was the Vickers Valiant, of which the prototype (*WB 210*) first flew on 18 May 1951. Two Mark 1 and one Mark 2 prototypes were built, and were followed by 104 production aircraft, the first of which (*WP199*) flew on 21 December 1953. They were powered by various versions of the Rolls-Royce Avon axial-flow turbojet, four

such engines being located in the wing roots. The Valiant entered RAF service with No 138 Squadron at Gaydon, Warwickshire, early in 1955 and afterwards equipped Nos 7, 49, 90, 148, 207, 214 and 543 Squadrons. The production also included versions for photo-reconnaissance (the B (PR) Mark 1) and tankers (B (PR) K Mark 1 and BK Mark 1). Maximum speed was 567 mph (912 km/h) at 36 000 ft (11 000 m) (Mach 0·84). Normal loaded weight with a 10 000 lb (4540 kg) bomb-load was 140 000 lb (63 560 kg). Range without external fuel tanks was 3450 miles (5550 km). A Valiant of Bomber Command carried Britain's first operational atomic bomb, which was dropped over Maralinga, Southern Australia, on 11 October 1956.

Supermarine Attacker of the Fleet Air Arm

Gloster Javelin Mk 1 all-weather fighters of No 46 Squadron, first to operate the world's first twin-jet delta-wing combat aircraft

The first jet pilot to achieve five confirmed aerial victories over jet aircraft was Captain James Jabara, an F-86 Sabre pilot of the 4th Fighter-Interceptor Wing, USAF, who shot down his fifth MiG-15 on 20 May 1951. Captain, later Major Jabara went on to destroy a total of 15 MiG-15s, thereby becoming the second most successful Allied pilot of the Korean War.

The first standardised jet fighter to serve in FAA first-line squadrons was the Supermarine Attacker, and was also the first aircraft powered by the Rolls-Royce Nene turbojet. The type first entered service with No 800 Squadron at Ford, Sussex, on 22 August 1951, and this was the FAA's first operational jet squadron. When withdrawn from first-line service the Attackers were transferred to RNVR air squadrons: when No 1831 Squadron received these aircraft on 14 May 1955, it became the first jet-fighter squadron of the RNVR.

The first British delta-wing interceptor fighter, and the first twin-jet delta fighter in the world, was the Gloster Javelin. First flight of the prototype (*WD804*) was made by Squadron Leader W A Waterton on 26 November 1951. The Javelin was also **the RAF's first purpose-built all-weather interceptor fighter.** The use of the delta wing posed many aerodynamic problems, early flights suffering from control surface vibration and buffeting. On the 99th flight, on 29 June 1952, both elevators were lost following violent flutter, and by superb flying Waterton managed to control the aircraft in pitch, by means of the variable incidence tailplane, and bring it in to a fast landing, which caused the landing gear to collapse. For his skill and courage in saving the aircraft and its flight recorder, Waterton was awarded the George Medal. **The first production Javelin FAW1** (*XA544*), powered by two 8150 lb (3697 kg) thrust Armstrong-Siddeley Sapphire AS Sa6 turbojet engines, made its first flight on 22 July 1954. First deliveries to No 46 Squadron at Odiham, Hants, began in February 1956.

The heaviest bomb-load carried by an operational bomber was that of the Boeing B-52 Stratofortress at 75 000 lb (34 019 kg); with this war-load on board, the B-52B possessed a range of approximately 3000 miles (4828 km). Dubbed 'the big stick', the YB-52 prototype was first flown on 15 April 1952 by A M 'Tex' Johnson. The first of three production B-52As was delivered to the USAF's Strategic Air Command (SAC) on 27 November 1957. B-52s subsequently became the main flying deterrent of SAC for twenty years.

Douglas A3D Skywarrior

The world's first large bomber to have a delta-wing plan-form was the Avro Vulcan B1, the prototype of which (*VX770*) flew for the first time on 30 August 1952. Production aircraft first entered service with RAF Bomber Command in May 1956, equipping No 230 Operational Conversion Unit at Waddington, Lincs.

The heaviest aeroplane ever to serve as standardised equipment on board aircraft carriers was the Douglas A3D Skywarrior carrier-based attack-bomber. The first of two XA3D-1 prototypes first flew on 28 October 1952, and initial production A3D-1s (later A-3A) were first delivered to VAH-1 (US Navy Heavy Attack Squadron One) on 31 March 1956. The definitive production version, ultimately designated A-3B, served on board carriers of the Essex and Midway classes. This latter version had a span of 72 ft 6 in (22·10 m) and a maximum loaded weight of 82 000 lb (37 195 kg). Powered by two 12 400 lb (5624 kg) thrust Pratt and Whitney J57-P-10 turbojets, the A-3B Skywarrior had a maximum level speed of 610 mph (982 km/h) at 10 000 ft (3050 m).

The most successful Allied fighter pilot of the Korean War was Captain Joseph McConnell, Jr, of the 16th Fighter Squadron, 51st Fighter-Interceptor Wing, USAF; an F-86 Sabre pilot, McConnell scored his 16th and last victory on 18 May 1953, a day on which he destroyed a total of three MiG-15s. He was subsequently killed testing a North American F-86H on 25 August 1954.

The first British transport aircraft specifically designed for air-dropping of heavy loads, and also the RAF's largest aircraft at the time of its introduction, was the Blackburn Beverley. Powered by four 2850 hp Bristol Centaurus 173 engines, the prototype (*WZ889*) flew for the first time on 14 June 1953. Able to carry a payload of almost 22 tons (22·4 tonnes), Beverleys began to equip No 47 Squadron Transport Command in March 1956.

The first American naval pilot to achieve five air victories over Korea was Lieutenant Guy Bordelon who, flying a piston-engined Vought F4U Corsair, shot down his fifth victim on 17 July 1953.

The only turboprop-powered flying-boat to serve with the US Navy was the Convair R3Y Tradewind, the first R3Y-1 flying for the first time on 25 February 1954. A large transport with a gross weight of 160 000 lb (72 570 kg), the later R3Y-2s had a nose loading door through which armoured vehicles could be disembarked directly on to a landing beach.

The first turbojet-powered all-weather fighter to serve with the Royal Navy was the de Havilland Sea Venom, which first entered service with No 890 Squadron, which reformed at Yeovilton, Somerset, on 20 March 1954. At the end of 1958 three Sea Venoms of No 893 Squadron carried out the first firings of Firestreak missiles by an operational fighter squadron of the Royal Navy.

The first supersonic operational carrier-borne naval interceptor in the world was the Grumman F11F-1 Tiger of the US Navy. Designated originally F9F-9, this was changed after the first three production aircraft had been delivered and, in 1962, was finally designated F-11. The prototype, powered by a Wright J65-W-6 turbojet rated at 7800 lb (3538 kg) thrust, flew for the first time on 30 July 1954. The F-11A was capable of a speed of 890 mph (1432 km/h) in level flight at 40 000 ft (12 200 m), and was armed with four 20 mm cannon. Two or four Sidewinder missiles could be carried on underwing pylons. F11F-1s entered service with the US Navy's VA-156 Squadron in March 1957. Two F11F-1Fs were powered by 15 000 lb (6805 kg) thrust J79-GE-3A engines, and demonstrated Mach 2 performance in level flight.

The first supersonic single-seat fighter to serve with the RAF, the English Electric (later BAC) Lightning, designed by W E W Petter, flew for the first time on 4 August 1954 piloted by Wg Cdr R P Beamont. Entering operational service with No 74 Squadron at Coltishall, Norfolk, in July 1960, it was the first RAF fighter capable of speeds in excess of Mach 2, and its first integrated weapons system. During the research which led to its construction, Britain's first transonic wind tunnel was built.

Blackburn Beverley heavy freighter

The first jet fighter in the world with a variable-incidence wing was the Chance Vought (subsequently Ling-Temco-Vought, or LTV) F-8 Crusader supersonic air-superiority fighter of the US Navy. The operation of such a high-performance fighter from the deck of an aircraft carrier meant that to achieve an acceptable landing speed an excessive nose-up attitude would result. The use of a variable-incidence wing provided the necessary compromise. The first of two XF8U-1 prototypes, powered by a Pratt and Whitney J57-P-11 turbojet engine, made its first flight on 25 March 1955. Deliveries of the production Crusader, under the designation F-8A, began to VF-32 (US Navy Fighter Squadron 32) on 25 March 1957, serving originally at sea aboard the USS *Saratoga*. This initial production version was powered by a J57-P-12 or -14 engine, providing a maximum level speed of 1100 mph (1770 km/h) at 40 000 ft (12 200 m). Armament comprised four 20 mm cannon and, on early models, a fuselage pack of 32 air-to-air rockets. Later F-8As carried two fuselage-mounted Sidewinder missiles.

The last British heavy bomber powered by piston engines was the Avro Lincoln four-engine aircraft. The Lincoln Mk 1, known originally as the Lancaster Mk IV, was powered by 1750 hp Rolls-Royce Merlin 85 engines. This version had a maximum level speed of 319 mph (513 km/h) at 18 500 ft (5640 m), could carry 14 000 lb (6350 kg) of bombs and had defensive armament of six 0·50 in machine-guns in pairs in nose, dorsal and tail turrets. The last Lincoln was retired from Bomber Command in December 1955, when the RAF Command became an all-jet force.

The world's first known air-transportable hydrogen bomb was dropped on 21 May 1956, from a Boeing B-52B flying at 50 000 ft (15 240 m) over Bikini Atoll in the Pacific Ocean.

The first turbojet aircraft in the world to be used in the military transport role was a modified version of the de Havilland Comet Series 2, the first of which was delivered to RAF Transport Command at Lyneham, Wilts, on 7 July 1956.

The first British atomic bomb was dropped by a Vickers Valiant, *WZ366*, of No 49 (Bomber) Squadron, captained by Squadron Leader E J G Flavell, AFC, over Maralinga, Southern Australia, on 11 October 1956.

The first British hydrogen bomb was dropped by a Vickers Valiant of No 49 (Bomber) Squadron, captained by Wing Commander K G Hubbard, OBE, DFC, AFC, on 15 May 1957. The bomb was detonated at medium altitude over the Pacific in the Christmas Island area.

The first swept-wing single-seat fighter to be built for the Royal Navy, and the first to be capable of low-level attack at supersonic speed, attained in a shallow dive, was the Supermarine Scimitar. The first operational squadron, No 803, was formed at Lossiemouth, Scotland, in June 1958. The Scimitar was also the first British naval aircraft to have a power-operated control system.

The first specially designed anti-submarine helicopter ordered for the Royal Navy was the Westland Wessex, developed from the Sikorsky S-58. Equipped with an automatic pilot, the Wessex could be operated by day or night in all weathers. Earliest first-line squadron to be equipped with the type was No 815, commissioned at Culdrose on 4 July 1961.

The world's first specially designed low-level strike aircraft was the Blackburn NA39, subsequently named Buccaneer. The S1, powered by de Havilland Gyron Junior turbojets, first entered operational service with the FAA's No 801 Squadron at Lossiemouth in July 1962. The developed S2 version, with more powerful Rolls-Royce Spey turbojets, also entered service with No 801 Squadron, on 14 October 1965. Prior to that, on 4 October 1965, the first production aircraft (*XN974*) became the first FAA aircraft to make a non-stop crossing of the North Atlantic without flight refuelling. The 1950 miles (3138 km) from Goose Bay, Labrador, to RNAS Lossiemouth were flown in 4 h 16 min.

The first pilotless anti-submarine helicopter to enter service with the US Navy was the Gyrodyne QH-50A. Entering service in 1963, the Navy's drone helicopters each carried two homing torpedoes.

The first Fleet Air Arm helicopter used extensively from platforms on frigates, and smaller vessels, was the Westland Wasp. The first Small Ship Flight was formed on 11 November 1963. Though small, the Wasp could pack a hefty punch, carrying two homing torpedoes, or depth charges or air-to-surface missiles.

The last operational flying-boat in US Navy service, retired from first-line duties in 1966, was the Martin P5M Marlin. The first production P5M made its first flight on 22 June 1951, and the type saw extensive use for maritime patrol, ASW and air-sea rescue.

The first supersonic (in level flight) carrier-based interceptor fighter to serve with the Royal Navy, the McDonnell Douglas F-4K Phantom (RN designation FG1), first entered service at Yeovilton in April 1968. Powered by two Rolls-Royce Spey turbofan engines, the Navy's Phantoms have a maximum speed of Mach 2·1 and carry eight air-to-air missiles for interception duties.

The first land-based pure-jet aircraft in the world to be built for anti-submarine duties and long-range maritime patrol, the Hawker Siddeley Nimrod, developed from the Comet 4C airliner, first entered service with RAF Strike Command's No 201 Squadron on 2 October 1969. Powered by four Rolls-Royce Spey turbofans, the Nimrod has a ferry range of 5755 miles (9265 km).

First McDonnell F-4K Phantom for the Royal Navy

The US Navy's first helicopter mine countermeasures (MCM) squadron, HM-12, was established late in 1970. A special version of the Sikorsky S-65 twin-turbine helicopter has been developed, with dual hydraulically-powered winches to stream and retrieve the tow, and the first of these RH-53D helicopters was delivered to squadron HM-12 in September 1973. The special equipment deployed by the tow is designed to destroy magnetic, acoustic or mechanical mines.

The longest association of a navigator and a pilot must be that recorded by William G Crooks, DFC, who retired from regular communication flights for Hawker Siddeley Aviation on 30 June 1974. After service in the RAF from March 1942, he first crewed with pilot R J Chandler in April 1948. From then, until his retirement, William Crooks recorded 6554 h as navigator/radio operator on flights with R J Chandler.

The largest military flying-boat in service today is the Japanese Shin Meiwa SS-2. Operated by the Japanese Maritime Self-Defence Force (JMSDF) under the designations PS-1 and US-1, signifying anti-submarine flying-boat and air-sea rescue amphibian respectively, the prototype of this four-turboprop aircraft made its first flight in the PS-1 configuration on 5 October 1967. With a wing span of 108 ft 8¾ in (33·14 m), the PS-1 has a maximum take-off weight of 94 800 lb (43 000 kg) and maximum ferry range of 2948 miles (4744 km).

Largest military flying-boat in current service, the Japanese Shin Meiwa PS-1

ORIGINS OF THE WORLD'S AIR FORCES AND CURRENT EQUIPMENT

Where numbers of aircraft are quoted, it should be realised that the figures are approximate. This applies especially to Eastern bloc countries where recent build-ups may have enlarged the air forces considerably.

Abu Dhabi The Air Wing of the Abu Dhabi Defence Force was formed with British help in 1968, following a UK decision to withdraw from the Persian Gulf area in 1971. Renamed Abu Dhabi Air Force in 1972. Current equipment includes Dassault Mirage 5s; Hawker Hunters; Agusta-Bell 205, Aérospatiale Alouette III and Puma helicopters; about eight transports and a small number of trainers. Strength: 20 combat aircraft.

Afghanistan The Royal Afghan Air Force was formed in 1924 by King Amanullah with an initial equipment of two Bristol F2B fighters flown by German pilots. Current equipment includes MiG-17s, MiG-19s, MiG-21s; Sukhoi Su-7s; Ilyushin Il-28s; Mil Mi-4 and Mi-8 helicopters; about 40 transports and a number of trainers. Strength: 150 combat aircraft, 9000 personnel.

Albania The Albanian Air Force was established in 1947 with initial equipment of 12 Yak-3 fighters. Current equipment, much of it supplied by China, includes MiG-15s, MiG-17s, MiG-19s, MiG-21s; Mil Mi-1 and Mi-4 helicopters; a number of transports and trainers. Strength: 75 combat aircraft, 6000 personnel.

Algeria Force Aérienne Algérienne was formed in 1962, with Egyptian and Soviet assistance, following attainment of Algerian independence from France. Current equipment includes MiG-15s, MiG-17s, MiG-21s; Sukhoi Su-7s; Ilyushin Il-28s; Aérospatiale Puma, Mil Mi-4, Mi-6 and Mi-8 and Hughes Model 269A helicopters; about 27 transports and many trainers (including 26 armed Magister trainers). Strength: 175 combat aircraft, 4500 personnel.

Angola An air force was formed in 1976 following the left-wing MPLA victory in the civil war, accomplished with Soviet and Cuban assistance. Current equipment includes MiG-15s, MiG-17s, MiG-21s; Fiat G91Rs; Aérospatiale Alouette III helicopters and several transports. Strength: 20 combat aircraft.

Mirage IIID of the Argentine Air Force

F-111 of the Royal Australian Air Force, taking off
with swing-wings in forward position (Herald and
Weekly Times, Melbourne)

Argentina The Argentine Air Force was
formed on 8 September 1912 with the establish-
ment of the Escuela de Aviación Militar at El
Palomar. The Argentine Naval Aviation Service
was formed on 17 October 1919 with the pre-
sentation by the Italian Government of facilities
established by an Italian mission at San Fernando.
Current equipment includes North American
F-86 Sabres; Dassault Mirage IIIs; McDonnell
Douglas Skyhawks; English Electric Canberras;
FMA Pucaras; Bell 47G and Iroquois, Fairchild-
Hiller UH-23, Hughes OH-6A and Sikorsky
S-55 helicopters; over 100 transports, and about
86 trainers and other miscellaneous types.
Strength: 120 combat aircraft, 18 000 personnel.
Army and Navy operate a further 150 aircraft,
including 50 naval combat aircraft, and the
military airline about 22 transports.

Australia The origin of Australian military
aviation is not well documented. It is known
that Army Order No 132 of 1912 created a single
flight to form a training school. This flight, comprising four officers, seven warrant officers and
sergeants and 32 air mechanics, was the first unit
of the Aviation Corps. The Australian Flying
Corps dates from approximately 27 December
1915, but this was abolished on 31 March 1921
when the Australian Air Force was created. On
13 August 1921 this service gained the prefix
Royal, and the Royal Australian Air Force
(RAAF) it has remained to this day. Current
equipment includes General Dynamics F-111s;
Dassault Mirage IIIs; BAC Canberras; Lockheed Neptunes and Orions; Bell Iroquois and
Boeing-Vertol Chinook helicopters; about 57
transports and about 150 trainers: Strength: 100
combat aircraft, 21 500 personnel. The Navy
operates over 100 aircraft, and the Army operates about 112 aircraft.

Austria Formation of the Deutschösterreich-
ische Fliegertruppe ('Austro-German Flying
Troop') took place on 6 December 1918. The
current Austrian Air Force (Österreichische Luftstreitkräfte) was founded in 1955 with four Yak-
11 and four Yak-18 trainers. Current equipment
includes SAAB-Scania 105s; Agusta-Bell 204B
and 206A, Bell Kiowa, Aérospatiale Alouette III and Sikorsky S-65 helicopters; about 14
transports and 15 trainers. Strength: 36 combat
aircraft, 4300 personnel.

Bangladesh Bangladesh Defence Force (Air
Wing) was established with Indian assistance
before the outbreak of hostilities in December
1971. Initial equipment of some half-dozen aircraft included DHC-3 Otters. Current equipment includes MiG-21s; Aérospatiale Alouette
III, Bell 212, Mil Mi-8 and Westland Wessex
helicopters; about 11 transports and a small
number of trainers. Strength: nine combat aircraft, 3000 personnel.

Belgium La Force Aérienne Belge came into
effective being on 5 March 1911 with the inauguration of its first airfield at Brasschaet, Antwerp. Current equipment includes Lockheed
Starfighters; Dassault Mirage 5s; Aérospatiale
Alouette II and III, and Westland Sea King helicopters; about 30 transports and over 100 trainers. On order are over 100 General Dynamics
F-16 fighters and many Alpha Jet trainers.
Strength: 140 combat aircraft, 20 000 personnel.

Bolivia The Cuerpo de Aviación was founded
in August 1924, although a Flying School had
been established at Alto La Paz as early as 1915.
Current equipment includes North American
F-86 Sabres; F-51 Mustangs; T-6s and T-28s;
Embraer Xavantes; Canadair T-33s; Hiller OH-
23 and Hughes 500 helicopters; about 40 transports and over 30 trainers. Strength: 50 combat
aircraft, 4000 personnel.

Brazil The Brazilian Naval Air Force was
founded with the establishment of a seaplane
school in 1913 at Rio de Janeiro with three American Curtiss seaplanes. The Army Air Service
was founded under French training supervision
in 1919 at Rio de Janeiro. Current equipment
includes Dassault Mirage IIIs; Lockheed AT-33s;
Northrop F-5Es; Embraer Xavantes; North
American T-6Gs; Neiva T-25 Universals; Lockheed Neptunes and Grumman Trackers; Kawasaki-Bell KH-4, Hughes OH-6A and Bell H-
13J, Iroquois, JetRanger and Model 47 helicopters; plus transports, trainers and other aircraft. Strength: 170 combat aircraft, 35 000
personnel.

Brunei Sultanate Only former British de-
pendency populated by Malay people that did

Below: This SR-71A 'Blackbird' strategic reconnaissance aircraft of the USAF set a new world air speed record of 2193.17 mph (3529.56 km/h) in July 1976.

Right: Three-nation 'swing-wing' combat aircraft, the multi-role Tornado will equip the air forces of the UK, Italy and West Germany in the 'eighties.

Below: The US Navy's nuclear aircraft carrier *Enterprise*, with its squadrons deployed on deck. On the stern are Grumman F-14A Tomcat 'swing-wing' fighters.

Left: Westland Lynx HAS. Mk 2 anti-submarine helicopter undergoing deck landing trials on HMS *Sheffield*.

Centre: Bell Model 214B BigLifter operating in the flying crane role.

Left: Mil Mi-8, standard medium-size helicopter of the Soviet airline Aeroflot and the Soviet armed forces (*Martin Fricke*).

Top: Boeing 747SP, short-fuselage long-range airliner of South African Airways.

Above: Another product of international co-operation — the A.300B European Airbus.

Right: Concorde in British Airways livery.

Top: Spaceship *Enterprise*, first US
Space Shuttle Orbiter designed to
replace expendable rockets as a
satellite launch vehicle.

Centre: Designed as a strategic bomber
to replace the B-52 Stratofortress, the
XB-70A Valkyrie was flown for the first
time on 21 September 1964. Only two
were built and used as aerodynamic
test aircraft able to fly at three times
the speed of sound. In this view the
wing tips are folded down to improve
stability and manoeuvrability.

Right: For research at low altitude —
the Bell XV-15 tilt-rotor research
aircraft with its engine pods and
rotor/propellers in an intermediate
position. The engines are tilted to a
vertical position for take-off and
landing, and to a horizontal position
for cruising flight.

not join the Federation of Malaysia in 1963; an Air Wing of the Royal Brunei Malay Regiment was formed in 1965. Initial equipment comprised three Westland Whirlwind Mk 10 helicopters. Current equipment includes Bell Iroquois, Kiowa and 212 helicopters, and a Hawker Siddeley 748 transport.

Bulgaria A Bulgarian Army Aviation Corps was formed originally in 1912 with Blériot and Bristol monoplanes and fought in the Balkan War of 1912–13. It was resurrected shortly after 12 October 1915 when Bulgaria entered the war as one of the Central Powers. Current equipment includes MiG-17s, MiG-19s, MiG-21s; Ilyushin Il-28s; Mil Mi-1, Mi-4 and Mi-6 helicopters; about 30 transports and a number of trainers. Strength: 250 combat aircraft, 25 000 personnel.

Burma Union of Burma Air Force is a small independent air force started with British aid in 1955. Initial equipment: 20 Sea Fury FB11s and some Vampire T55s. Current equipment includes: North American F-86 Sabres; Lockheed AT-33s; Kawasaki-Bell KV-107-II and Model 47, Kaman Huskie, Aérospatiale Alouette III and Bell Iroquois helicopters; plus about 22 transports and about 40 trainers. Strength: 10 combat aircraft, 7000 personnel.

Cambodia   see **Kampuchea**

Cameroon A small Cameroon Air Force was established by the Cameroon Republic during the 1960s. Current equipment includes Aérospatiale Alouette II, III and Puma helicopters, nine transports and eight other aircraft. Strength: 300 personnel.

Canada During the First World War, many Canadians served in the British RNAS, RFC and RAF. Late in 1918, in Britain, an embryo 'Canadian Air Force' was represented by two Canadian squadrons, but the Armistice virtually ended its existence. In 1919 an Air Board was set up to control a 'non-permanent militia' basis of part-time flying instruction, and on 19 April 1920 this board began administration of a Canadian Air Force, established with considerable assistance from the RAF. On 1 April 1924 the Royal Canadian Air Force (RCAF) came into being. It is now Air Command of the unified Armed Forces. Current equipment includes Canadair CF-104G Starfighters and CF-5As; McDonnell Douglas Voodoos; Canadair Argus; Grumman Trackers; Avro Canada CF-100s; Bell Iroquois, Kiowa, Twin-Pac, Boeing Vertol Chinook and Voyageur, and Sikorsky Sea King helicopters; over 100 transports, about 340 trainers and various other types. Strength: 225 combat aircraft, 23 000 personnel.

Central African Republic As a result of a bilateral defence agreement with France, the Force Aérienne Centrafricaine was established in 1960. Current equipment includes Aérospatiale Alouette II, Bell 47G and Sikorsky H-34 helicopters; five transports and 17 other aircraft. Strength: 100 personnel.

Chad A small air force was set up with help from France. Current equipment includes Douglas Skyraiders; Aérospatiale Alouette II and III and Sikorsky H-34 helicopters; about 16 transports and a few other types. Strength: five combat aircraft, 300 personnel.

Chile The Chilean Air Force was formed with the establishment of a flying school at Lo Espejo on 11 February 1913. Current equipment includes Northrop F-5Es; Cessna A-37Bs; Hawker Hunters; Douglas Invaders; Neiva T-25 Universals; Grumman Albatross; Bell Iroquois, Hiller UH-12E, Sikorsky S-55T and UH-19 helicopters; about 58 transports and 100 trainers and various other types. Strength: 75 combat aircraft, 10 000 personnel.

China The Chinese Army Air Arm came into being with the establishment in January 1914 of a flying school at Nan Yuan at which members of the Chinese Army commenced flying instruction under an American, Art Lym. Current equipment includes about 4100 MiG-17s, MiG-19s and MiG-21s; 55 Tupolev Tu-16 strategic bombers; 300 Ilyushin Il-28s; over 300 Mil and other helicopters; about 400 transports and other aircraft. Naval aviation includes about 100 Ilyushin Il-28s; Beriev Be-6s; helicopters.

Colombia The Fuerza Aérea Colombiana originated in the Escuela de Aviación (flying school) which was founded at Flandes with a Caudron G IIIA-2 aircraft on 4 April 1922, on the authority of the Colombian Minister of War, Dr Aristóbulo Archila. Current equipment includes Dassault Mirage 5s; Douglas Invaders; Lockheed AT-33s; Bell 47, Iroquois and 212, Hughes OH-6 and TH-55, Hiller OH-23 and Kaman HH-43B Huskie helicopters; over 70 transports and over 800 trainers. Strength: 30 combat aircraft, 6000 personnel.

Costa Rica This air force is known as the Guardia Civil Air Wing and is basically an extension of the Ministry of Public Security. Current equipment includes Sikorsky S-58 and Fairchild-Hiller FH-1100 helicopters, three de Havilland Otters and four Cessna liaison aircraft.

Cuba Following the Castro revolution in

1958, the Cuban Revolutionary Air Force was modernised and expanded with the assistance of Eastern bloc countries. Initial re-equipment included MiG-17s, MiG-19s, Il-14s and An-2s. Current equipment includes MiG-15s, MiG-17s, MiG-19s, MiG-21s; Mil and other helicopters; about 50 transports and about 80 trainers.

Cyprus   The Air Wing of the Cyprus National Guard was formed after Cyprus became an independent republic on 16 August 1960. Initial equipment included some light aircraft and helicopters. Current equipment includes several Dornier, Beech and Piper light transport and liaison aircraft.

Czechoslovakia   The Czechoslovak Army Air Force was formed early in 1919 from air components previously serving with the Czech Legions in Russia and France. Current equipment includes MiG-17s, MiG-21s; Sukhoi Su-7s; about 50 transport aircraft; 100 helicopters; and 300 trainers. Strength: 450 combat aircraft, 45 000 personnel.

Denmark   The Danish Army Air Corps was established on 2 July 1912 with the formation of a flying school. Current equipment of the Royal Danish Air Force includes Lockheed and Canadair Starfighters; SAAB-Scania Drakens; North American F-100 Super Sabres; Aérospatiale Alouette III and Sikorsky S-61 helicopters; 15 transports and about 70 trainers. On order are General Dynamics F-16 fighters. Strength: 116 combat aircraft, 8500 personnel.

Dominican Republic   Current equipment includes refurbished North American F-51 Mustangs; de Havilland Vampires; Douglas Invaders; about 17 transport aircraft; 15 helicopters and 20 trainers. Strength: 30 combat aircraft, 3500 personnel.

Dubai   Current equipment of the Dubai Police Air Wing includes Aermacchi MB326s; SIAI-Marchetti SF-260s; Bell 205A and JetRanger helicopters. Strength: three combat aircraft.

Ecuador   Formation of the original Cuerpo de Aviadores Militares began in 1920 under the supervision of an Italian Aviation Mission. Current equipment of the Ecuadorean Air Force includes English Electric Canberras; Gloster Meteors; BAC Strikemasters; SEPECAT Jaguars; about 27 transports; 15 helicopters; 50 trainers. Strength: 35 combat aircraft, 2750 personnel.

Egypt   The Egyptian Army Air Force was originally planned in 1930 under British influence, but its official foundation was not effected until May 1932 with the arrival of its first five aircraft (Gipsy Moths) from Britain. Current equipment includes about 230 MiG-21s; 48 MiG-23s and -27s; 32 Dassault Mirage IIIs; over 100 MiG-17s; more than 100 Sukhoi Su-20s and Su-7s; 25 Tupolev Tu-16 strategic bombers; more than 170 helicopters; 75 transports; about 250 trainers; and other types. Strength 30000 personnel.

El Salvador   Current equipment includes 18 Dassault Ouragans; about 13 transports; one Fairchild-Hiller FH-1100 helicopter and 14 trainers. Strength: 1000 personnel.

Ethiopia   The Ethiopian Air Force was formed in 1924 with the procurement of French and German aircraft by Ras Tafari (later Emperor Haile Selassie). Current equipment includes Northrop F-5s; North American F-86 Sabres; BAC Canberras; Cessna A-37Bs; North American T-28Ds; about 31 helicopters; 38 transports; about 35 trainers and other types. Strength: 55 combat aircraft, 2400 personnel.

Finland   Current equipment includes MiG-21s; SAAB-Scania Drakens; Aérospatiale Alouette II, Agusta-Bell 206A, Hughes 500M, Mil Mi-4 and Mi-8 helicopters; 10 transports; about 90 trainers and other types.

France   The French Air Force (originally the Service Aéronautique) was founded as a separate command in April 1910. By this time several army pilots had learned to fly and the new command had been issued with a Blériot, two Wrights and two Farmans. Current equipment includes over 190 Dassault Mirage IIIC/Es; 115 Dassault Mirage F1s; 48 Dassault Mirage M-5Fs; 160 SEPECAT Jaguars; 52 North American F-100 Super Sabres; 50 Dassault Mirage IVAs; over 30 Aérospatiale Vautours; nearly 60 Dassault Mirage III reconnaissance aircraft; about 200 transports; many trainers; about 135 helicopters and many other types. Dassault Mirage 2000s on order. Strength: 600 combat aircraft,

MBB 320 Hansa of the Luftwaffe, equipped for ECM duties (Ing. Hans Redemann)

18 silo-based strategic nuclear missiles, 102 000 personnel. Large Navy and Army air arms.

Gabon Formed in 1961, with French assistance, the Air Force has six Dassault Mirage 5s; ten transports and seven helicopters. Strength: 100 personnel.

Germany (West) The original German Military Air Force owed its origin to the purchase by the Army of its first Zeppelin dirigible in 1907 and its first 11 aeroplanes in 1910. The Military Aviation Service was formally established on 1 October 1912. Current equipment includes McDonnell Douglas Phantom IIs (including reconnaissance); Lockheed Starfighters; Fiat G-91s; 120 Bell Iroquois and Aérospatiale Alouette II helicopters; about 90 transports; 300 trainers, and other types. On order are Panavia Tornados and Alpha Jets. Strength: 460 combat aircraft, 110 000 personnel. Navy has about 200 aircraft (about 110 of which are Starfighters), with Panavia Tornados on order. Army has over 500 helicopters, with large number of MBB Bo 105s on order.

Germany (East) Organised on tactical lines for army support, under the East German Defence Ministry, the air section of the Nationale Volksarmee, known as the Air Force of the German Democratic Republic, originated in 1950 as a branch of the Volkspolizei. Current equipment includes about 200 MiG-21s; 90 Sukhoi Su-7s; 50 MiG-17s; about 80 Mil helicopters; transports; and trainers. Strength: 35 000 personnel. Soviet aircraft based in East Germany include the latest MiG and Sukhoi fighters and reconnaissance aircraft and Mil Mi-24 heavy helicopter gunships. East German Army and Navy have many Mil helicopters.

Ghana The Ghana Air Force was formed with the help of instructors from India and Israel and was equipped initially with Hindustan HT-2 basic trainers. In late 1960 RAF ground and flying personnel began to provide instruction and the first student pilots were sent to the UK for training. Current equipment includes 20 transports, six helicopters and 19 trainers. Strength: 1250 personnel.

Great Britain The Royal Air Force owes its origins to the balloon experiments by the Royal Engineers at Woolwich which commenced in 1878. The Air Battalion of the RE was established in February 1911, and the Royal Flying Corps in May 1912 with Military and Naval Wings. In November 1913 the Admiralty announced the formation of the Royal Naval Air Service. The RFC and RNAS continued as separate services until amalgamated to form the Royal Air Force on 1 April 1918. The Fleet Air Arm came into being in 1924. Current RAF equipment includes between 30 and 50 Avro Vulcans; about 60 Hawker Siddeley Buccaneers; 50 BAC Lightnings; 100 McDonnell Douglas Phantom IIs; 50 Hawker Siddeley Harriers; 100 SEPECAT Jaguars; 60 Hawker Hunters; 49 Hawker Siddeley Nimrods; 60 BAC Canberras; 12 Avro Shackletons for airborne early warning (to be replaced by Nimrods); nine Hawker Siddeley Argosys; over 20 Handley Page Victor tankers; 67 transports; about 150 helicopters and about 500 trainers. On order are 385 Panavia Tornados. Strength: 87 000 personnel. Royal Navy operates a few McDonnell Douglas Phantom IIs; Hawker Siddeley Buccaneers; many helicopters; a few Westland Gannet early warning aircraft; trainers and transports. Army operates between 450 and 500 helicopters, plus about 50 other aircraft.

Greece The Royal Hellenic Army formed its first military squadron of four Farman biplanes in September 1912 at Larissa, its pilots being trained in France. In February 1914 the Naval Air Service was established under the guidance of a British Naval Mission. Current equipment of the Hellenic Air Force includes Dassault Mirage F1s; McDonnell Douglas Phantom IIs; Vought Corsair IIs; Northrop F-5As/RF-5As; Lockheed F-104G Starfighters; Rockwell T-2 Buckeyes; Grumman HU-16 Albatross; 75 transports; 40 helicopters; more than 100 trainers. Strength: 275 combat aircraft, 22 000 personnel.

Guatemala Military aeroplanes were first flown by the Guatemalan Army shortly after the First World War, but it was not until 1929 that the Cuerpo de Aeronautica Militar was established as an echelon of the Army. Current equipment of the Guatemalan Air Force includes 8 Cessna A-37Bs; about 17 transports; ten heli-

Panavia Tornado multi-role combat aircraft, under development for the Royal Air Force and the air forces of Germany and Italy

Post War—Military continued on page 170

Gloster Gladiator II *flown late in 1940 by Flight-Lieutenant M T St J Pattle, DFC, as commander of 'B' Flight, No 80 (Fighter) Squadron, RAF; the serial number of this aircraft is believed to have been K7971. Flying from various bases in Greece, Pattle is known to have shot down at least twenty-four enemy aircraft by the end of 1940. Converting on to Hawker Hurricanes early in 1941, he went on to shoot down an estimated total of forty-one enemy machines before his death in action over the Piraeus on 20 April 1941. The South African-born Pattle was thus the most successful fighter pilot to serve with the British air forces during the Second World War.*

Supermarine Spitfire IX, *serial EN398, flown by Wing Commander James E. ('Johnnie') Johnson, DFC, in April 1943 as Wing Leader of the Kenley Wing – later No 127 Wing, RAF. When Johnson took command of the Wing in March 1943 his score of confirmed victories was six; by the time he relinquished command in September 1943, it stood at twenty-five. Johnson went on to achieve thirty-eight confirmed aerial victories, all single-engined German fighters, and was thus the most successful English fighter pilot of the Second World War, the second most successful RAF fighter pilot, and the most successful Allied fighter pilot in terms of single-engined enemy fighters destroyed in aerial combat.*

Lavochkin La-5FN *flown between May and mid July 1944 by Captain Ivan N Kozhedub, Hero of the Soviet Union; he is believed to have been operating in the Ukraine during this period. By the close of hostilities Kozhedub had attained the rank of Guards Lieutenant-Colonel, and had been awarded the Gold Star of a Hero of the Soviet Union three times. His final score of aerial victories is stated to be sixty-two which qualifies him as the most successful Allied fighter pilot of the Second World War.*

Messerschmitt Bf 109F-4/Trop, *5237, flown during June 1942 by Oberleutnant Hans-Joachim Marseille, Staffelkapitän of 3 Staffel, Jagdgeschwader 27. Based at Ain-El Gazala in Libya, Marseille scored his 101st confirmed aerial victory on 18 June 1942, and was subsequently awarded the Knight's Cross with Oak Leaves, Swords and Diamonds on 3 September 1942. His final score of 158 victories qualifies him as the most successful German fighter pilot to see combat exclusively against the British and Commonwealth air forces during the Second World War.*

Messerschmitt Bf 109G-14 *flown in February 1945 by Major Erich Hartmann as commanding officer of II Gruppe, Jagdgeschwader 52. During his service on the Russian Front Hartmann achieved a total of 352 confirmed aerial victories, and is thus the most successful fighter pilot the world has ever known. Fuller biographical details may be found in the body of the text.*

Junkers Ju 87D-5 *flown during the winter of 1943/44 by Major Hans-Ulrich Rudel as commanding officer of III Gruppe, Schlachtgeschwader 2 'Immelmann' in Russia. The most successful of Germany's Stuka pilots, and probably the greatest ground-attack pilot of that or any other war, Rudel flew a total of 2530 combat sorties, destroyed 519 Soviet armoured vehicles, and was the only man ever awarded (on 1 January 1945) the Golden Oak Leaves to the Knight's Cross.*

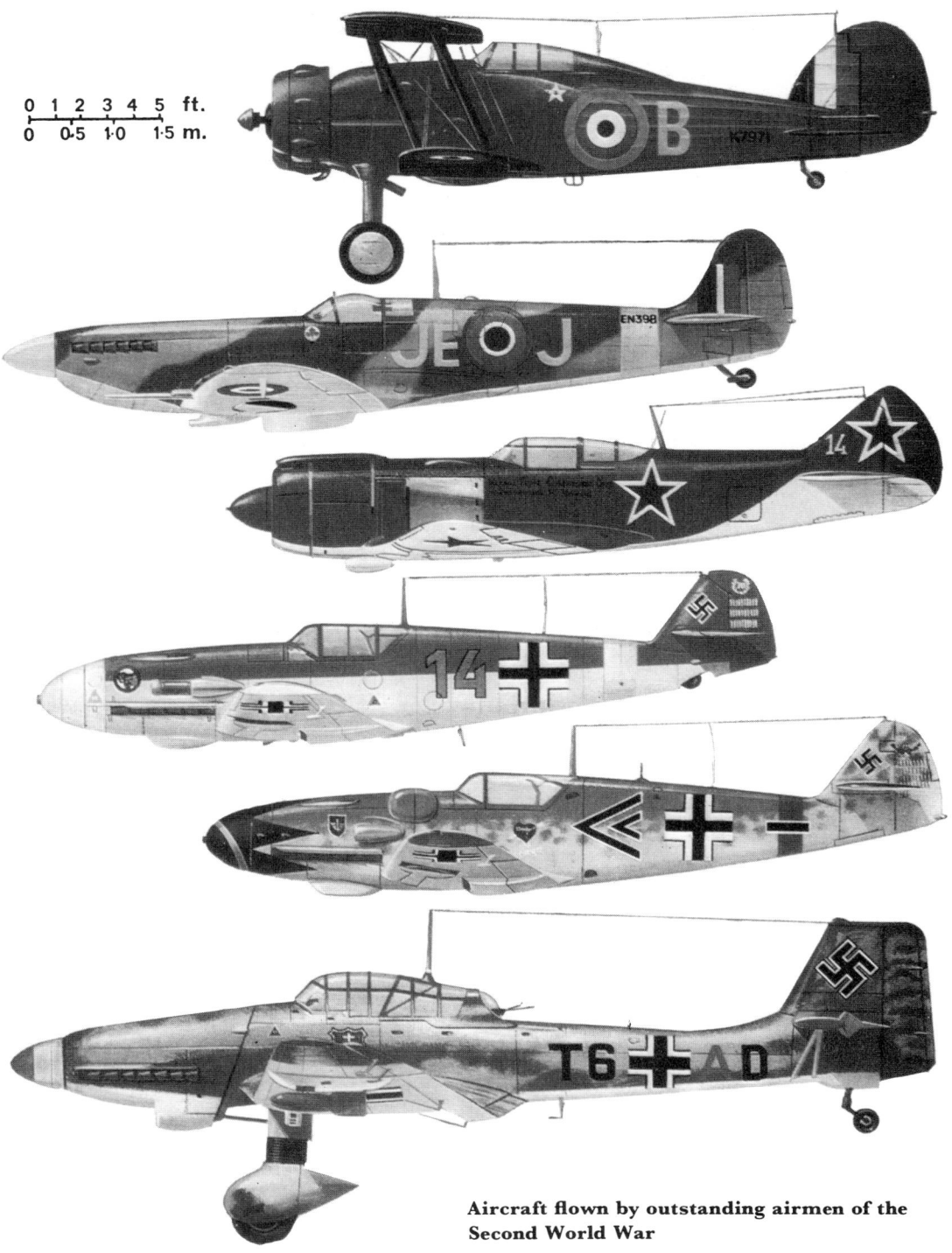

0 1 2 3 4 5 ft.
0 0.5 1.0 1.5 m.

**Aircraft flown by outstanding airmen of the
Second World War**

copters and 15 trainers. Strength: about 1000 personnel.

Republic of Guinea The Guinea Air Force was formed with assistance from the Communist bloc, following break with France in 1958. Current equipment includes eight MiG-17s; ten transports; two helicopters and a few trainers. Strength: 800 personnel.

Guyana This former British colony has a small air force that became operational in 1968. Current equipment includes three Britten-Norman Islanders, two Helio Couriers and two Hughes 269 helicopters.

Haiti The Haitian Corps d'Aviation was formed in 1943, primarily as a national mail-carrying organisation; in the late 1940s air patrols were added to its duties. Current equipment includes a few North American F-51 Mustangs; seven Sikorsky S-55 and H-34 helicopters; several transports; five trainers and seven Cessna liaison aircraft.

Honduras Current equipment of the Honduran Air Force includes Dassault Super Mystères; Cessna A-37Bs; North American F-86K Sabres; Lockheed RT-33s; North American T-28Es; and small numbers of transports, helicopters and trainers. Strength: 18 combat aircraft, 1500 personnel.

Hong Kong Crown Colony Formed with RAF assistance, the Royal Hong Kong Auxiliary Air Force has been operating since 1 May 1949. Initial equipment comprised some Harvard trainers and Spitfires. Current equipment includes three Aérospatiale Alouette III helicopters, one Britten-Norman Islander and a Beech Musketeer.

Hungary Forbidden under the terms of the Treaty of Versailles, Hungary as a separate Republic did not make provision for a small air service until 1936 when it made limited purchases of German and Italian aircraft. Current equipment of the air force includes MiG-21s; MiG-17s; Sukhoi Su-7s; and strong transport, helicopter and training elements. Strength: 150 combat aircraft, 20000 personnel.

India Scene of many years of British aviation influence, India established its own air force on 1 April 1933 with one squadron of Westland Wapiti general-purpose aircraft. Current equipment includes MiG-21s; HAL Gnat F1s; Hawker Hunters; Sukhoi Su-7Bs; BAC Canberras; HAL Maruts; about 260 transports; about 380 helicopters; numerous trainers and other types. Strength: 700 combat aircraft; 100000 personnel. Navy has about 25 Hawker Siddeley Sea Hawks, five Breguet Alizés, 3 Ilyushin Il-38s; 36 helicopters, a few transports, reconnaissance aircraft and trainers.

Indonesia After transfer of sovereignty from the Dutch to the United States of Indonesia on 27 December 1949, the Netherlands assisted in the establishment of the Indonesian Republic Air Force (Angkatan Udara Republik Indonesia) which, in the following year, took over from the Netherlands Indies Air Force, which was disbanded. Current equipment includes Commonwealth CA-27 Sabres; North American F-51D Mustangs; Rockwell OV-10F Broncos; Douglas Invaders; about 20 helicopters; 50 transports; trainers and liaison aircraft; six Grumman Albatross coastal patrol aircraft and other types. Strength: 60 combat aircraft, 28000 personnel. Army operates six helicopters, four liaison aircraft and seven transports. Navy operates five Grumman Albatross, 15 transports and several helicopters.

Iran Aviation in Iran originated in the Air Department of the Army Headquarters established in 1922 by the Prime Minister, Reza Khan. The first military aircraft was a Junkers F-13 transport based at Galeh-Morghi. Current equipment of the Imperial Iranian Air Force in-

Mikoyan MiG-21M licence-built by Hindustan Aeronautics for the Indian Air Force

Grumman F-14A Tomcat variable-geometry fighter of the Imperial Iranian Air Force

cludes Grumman F-14 Tomcats; McDonnell Douglas Phantom IIs; Northrop F-5s; Lockheed Orions; a number of helicopters; over 100 transports (including about 60 Hercules) and over 100 trainers. Strength: 500 combat aircraft, 60000 personnel. The Navy operates about 72 aeroplanes and helicopters, while the Army operates about 770 aircraft, mostly Bell helicopters, including 200 AH-1J gunships.

Iraq The Royal Iraqi Air Force was formed in 1931 with five de Havilland DH60T Gipsy Moths flown by Cranwell-trained Iraqi pilots. Current Iraqi Air Force equipment includes MiG-23s; MiG-21s; MiG-17s; Sukhoi Su-20s; Sukhoi Su-7Bs; Tupolev Tu-16s; Ilyushin Il-28s; Hawker Hunters; over 40 transports; about 115 Alouette III and Mil helicopters; and many trainers. Strength: 300 combat aircraft, 13000 personnel.

Ireland The Irish Army Air Corps was formed in 1922 after the completion of the Anglo-Irish Treaty of December 1921. Its first aeroplane was a Martinsyde Type A Mark II which had been purchased during the truce period to assist General Michael Collins to escape from England had the London talks failed. Current equipment includes SIAI-Marchetti SF-260W Warriors; Aérospatiale Super Magisters; Cessna FR172M patrol aircraft; eight Aérospatiale Alouette III helicopters; and 15 trainers. Strength: 24 armed aircraft, 600 personnel.

Israel The Israel Defence Force/Air Force owes its origin to the Sherut Avir, a military air service planned in 1947 at the time of Israel's emergence as a Sovereign State. The Sherut Avir gave way to the Israeli Air Force (Chel Ha'avir) in March 1948, and this in turn was integrated with the Army and Navy as the IDF/AF in 1951. Current equipment includes McDonnell Douglas F-15 Eagles, F-4 Phantoms and A-4 Skyhawks; Dassault Mirage IIIs; IAI Neshers and Kfirs; more than 40 transports, including Hercules and Boeing 707 dual-role tanker/transports; at least 150 helicopters; many trainers, including armed Magisters. Strength: over 500 combat aircraft, 17000 personnel.

Italy Successful use of aircraft by the Italian Army in the Italo-Turkish War of 1911–12 led to the formal establishment of the Air Battalion (Battaglione Aviatori) under the Ufficio d'Ispezione Servizi Aeronautica on 27 June 1912. A fully-fledged Military Aviation Service followed on 28 November 1912. Current equipment includes Lockheed F-104S, F-104G and RF-104G Starfighters; Aeritalia G-91Ys and

G-91Rs; Breguet Atlantics; Grumman Trackers; more than 150 ECM, transport and communications aircraft; more than 150 helicopters; over 250 trainers. Strength: 300 combat aircraft, 70000 personnel. The Naval Air Arm has about 70 helicopters. Army Aviation has more than 300 helicopters and 150 light aircraft.

Ivory Coast Republic Following independence, this former colony received military assistance from France for the establishment of its own armed forces. The Ivory Coast Air Force was formed following the receipt of one Douglas C-47 and two Broussard liaison aircraft in 1962. Current equipment includes eight Aérospatiale helicopters; five transports; six trainers and 11 other aircraft. Strength: 200 personnel.

Jamaica The Air Wing of the Jamaica Defence Force was established in July 1963, its initial equipment one Cessna 185 Skywagon and one regular pilot. Royal Canadian Air Force personnel trained flying and ground crews needed for subsequent expansion. Current equipment includes eight Bell helicopters; one Twin Otter and one Britten-Norman Islander; and six Cessna and Beech light aircraft.

Japan Origins of military aviation in Japan date back to July 1909 with the formation of the Temporary Military Balloon Research Committee. In 1911 the Army and Navy formed separate air services, the Japanese Army Air Force and the Imperial Japanese Naval Air Force. The current Japan Air Self Defence Force was formed on 1 July 1954. Current equipment includes Lockheed Starfighters; McDonnell Douglas F-4EJ and RF-4EJ Phantoms; North American F-86F and RF-86F Sabres, beginning to be replaced by Mitsubishi F-1 fighters; about 45 transports; over 350 trainers; 35 helicopters. Strength: 450 combat aircraft, 42000 personnel. Navy has about 350 aircraft, including 125 Lockheed Neptune and Kawasaki P-2J ASW aircraft, and 22 Shin Meiwa PS-1 ASW flyingboats. Army has about 450 aircraft, mostly helicopters.

Jordan The Royal Jordanian Air Force was formed in 1949 as the Arab Legion Air Force after the Arab–Israeli War of that year. Equipment was initially one de Havilland Rapide. Current equipment includes Northrop F-5As and F-5E Tiger IIs; a squadron of F-104A Starfighters; about 15 transports; 18 helicopters; 18 trainers. Strength: 70 combat aircraft, 6600 personnel.

Kampuchea When Cambodia gained independence from France in 1953, a small air force –

Aviation Nationale Khmere – was established with French assistance on 1 April 1954. Later, it expanded considerably with aircraft procured from both East and West. Current equipment includes Helio AU-24 Stallions; North American T-28Ds; about nine transports; 25 helicopters; and 20 more trainers. Strength: 30 armed aircraft.

Kenya The Kenya Air Force was established following that country's attainment of independence in 1963. Equipped originally with some ex-RAF de Havilland Chipmunks, it was inaugurated officially on 1 June 1964. Current equipment includes Hawker Hunters; BAC Strikemasters; 20 transports; five helicopters; five armed trainers. Strength: eight combat aircraft, 750 personnel.

Korea (North) With Soviet assistance the Korean People's Armed Forces Air Corps (KPAFAC) was formed in October 1948 to absorb the North Korean Army's Aviation Division which, using a small number of ex-Japanese Second World War aircraft had in turn originated in the North Korean Aviation Society in 1946. Current equipment includes Ilyushin Il-28s; MiG-21s, MiG-19s and MiG-17s; about 40 Mi-4 and Mi-8 helicopters; at least 40 transports; and trainers. Strength: 550 combat aircraft, 40000 personnel.

Korea (South) The Republic of Korea Air Force was established on a limited basis in 1949, the year before the North Korean forces crossed the 38th Parallel. The three-year war resulted in rapid expansion by means of massive assistance from the US Air Force. Current equipment includes McDonnell Douglas F-4D/E Phantoms; Northrop F-5As and F-5E Tiger IIs; Rockwell OV-10G Broncos; Northrop RF-5As; Hughes 500M-D Defender anti-tank helicopters; 15 other helicopters; 40 transports; 60 trainers. Strength: 300–350 combat aircraft, 30000 personnel.

Kuwait Originating as an air component formed in the 1950s, the Kuwait Air Force was established in 1960 following assistance given by a British advisory mission. It has undergone rapid expansion and current equipment includes Dassault Mirage F1s; BAC Lightnings; McDonnell Douglas Skyhawks; Hawker Hunters; BAC Strikemasters; six transports; and 35 helicopters. Strength: 75 combat aircraft.

Laos Military aviation in Laos originated in 1955, with French aid mainly for training purposes and the provision of some 27 aircraft from America. The Royal Lao Air Force was in-

augurated in August 1960, and expanded subsequently with assistance from the American MAP. At the time of the Pathet Lao take-over in 1975, about 70 North American T-28D and Douglas AC-47 counter-insurgency aircraft remained, with about 20 transports, over 40 helicopters and other types. It is not known which of these have been retained in service.

Lebanon The Lebanese Air Force was established in 1949 with RAF assistance, and was equipped initially with two Percival Prentice trainers. Equipment includes Hawker Hunters; Dassault Mirage IIIs; about 25 helicopters; one transport and several trainers. Strength: nominally 20 combat aircraft, 1000 personnel. Current status unknown.

Libyan Republic Military aviation in Libya originated with the receipt of two Gomhuria primary trainers, donated by the United Arab Republic in 1959. The original Royal Libyan Air Force was established in 1963, and became a small but efficient force, with US assistance. Following overthrow of the Monarchy on 1 September 1969, the Libyan Republic Air Force was created. Expansion has been rapid, with Soviet assistance. Current equipment includes MiG-23s and -27s; Dassault Mirage IIIs, III-Rs and 5s; about 17 transports; 30 helicopters; trainers and other types. Tupolev Tu-16 and Tu-22 bombers are based in Libya, some with LRAF markings. Strength: 130 combat aircraft, 5000 personel.

Malagasy Republic Madagascar's armed forces, built up with assistance and equipment from France, established the Malagasy Air Force on 24 April 1961 upon the receipt of seven aircraft, handed over from the French Air Force. Current equipment consists of about 30 transport and army support aircraft, including ten Douglas C-47s. Strength: 300 personnel.

Malaysia Origins of military aviation in the Federated Malay States date from the Straits Settlements Volunteer Air Force created in 1936. Although the RAF assumed the major share of operations against Communist terrorists, the Malayan Auxiliary Air Force was brought into being in 1950. From this was formed the Royal Malayan Air Force on 1 June 1958, renamed the Royal Malaysian Air Force following creation of Malaysia on 16 September 1963. Current equipment includes Northrop F-5E Tiger IIs; Canadair CL-41G Tebuans; about 50 helicopters; 30 transports, including six Lockheed C-130H Hercules; and 16 trainers. Strength: about 30 combat aircraft, 5250 personnel.

Four of the 101 Dornier
Do 28D-2s that have been
delivered to the Federal
German Luftwaffe for
general duties.

Three Socata Rallye light aircraft, built by a subsidiary of the French Aérospatiale company which developed the Concorde supersonic airliner with BAC of the UK.

The unique Australian-built Transavia Airtruk agricultural aircraft. Chemical loaders are able to drive right up to the fuselage hopper between the separate tailbooms.

Anglo-French Jaguar GR Mk 1 of No 54 Squadron, Royal Air Force.

Aeritalia-built Lockheed F-104S Starfighters of the Italian Air Force's 36th Stormo.

Ryan monoplane Spirit of St Louis *made one of the most celebrated flights of all time when Charles Lindbergh used it for the first solo non-stop transatlantic flight, from New York to Paris, thereby winning a $25,000 prize offered by Raymond Orteig. To make the 3610 mile (5810 km) flight possible, a huge fuel tank was fitted in front of the cabin, making it necessary to use a periscope to see forward, around the tank. Time taken was 33½ hours, on 20–21 May 1927. The Spirit of St Louis had a 233 hp Wright Whirlwind engine, span of 46 ft (14·02 m), length of 27 ft 6 in (8·38 m), weight of 5250 lb (2380 kg) at take-off and average speed of 108 mph (174 km/hr).*

Eight Handley Page H.P.42s *were built for service with Imperial Airways. First of them was G-AAGX, named* Hannibal, *which flew in November 1930 and became one of the four 18/24-passenger H.P.42Es used on the company's Cairo–Karachi–Cape Town routes. G-AAXC* Heracles *(illustrated) was first of the 38-seat H.P.42Ws (45s) used on European services from Croydon. Last of the big biplanes, these aircraft remained in use throughout the 'thirties and into the Second World War,* Heracles *alone logging 1¼ millon miles and carrying 95 000 passengers in its first seven years. Powered by four 550 hp Bristol Jupiter XFBM engines, it spanned 130 ft (39·62 m), had a length of 89 ft 9 in (27·36 m), loaded weight of 29 500 lb (13 380 kg) and maximum speed of 127 mph (204 km/hr).*

The Boeing 247, *although overshadowed by the later DC-2 and DC-3, was the airliner which set the pattern of 'modern' all-metal low-wing monoplane design, with retractable undercarriage, NACA-cowled engines driving controllable-pitch propellers, trim-tabs, a de-icing system and other innovations. Illustrated is NC 13347, which survives in the insignia of its original operator, United Air Lines. Seventy-two Boeing 247s were built, of which the first flew on 8 February 1933. Span was 74 ft (22·55 m), length 51 ft 4 in (15·65 m), loaded weight 12 650 lb (5738 kg) and maximum speed 182 mph (293 km/hr).*

The VS-300 helicopter, *first flown on 14 September 1939, was the world's first entirely-practical helicopter and established the now-conventional 'single-rotor' (one main rotor and one tail rotor) configuration. Designed and flown by the same Igor Sikorsky who, before he emigrated to America, had built the four-engined Bolshoi in Russia, the VS-300 had up to three tail rotors at one stage. Its fabric covering was not added until quite late in its test flying programme. Powered by a 75 hp Lycoming (later 100 hp Franklin) engine, the VS-300 was a single-seater with a rotor diameter of 30 ft (9·14 m), length of 27 ft 10 in (8·48 m), loaded weight of 1290 lb (585 kg) and speed of 40–50 mph (64–80 km/hr).*

The Heinkel He 178 *was the first jet-propelled aeroplane to fly, on 27 August 1939, although preceded by the rocket-powered He 176. Because of the imminence of the Second World War, the jet flight was kept secret. Nor did it afford much benefit to the Heinkel company, as Luftwaffe contracts for production jet-fighters went to its competitor, Messerschmitt. Power plant of the He 178 was a Heinkel He S3B turbojet designed by Pabst von Ohain, who ran his first engine only a short time after that of Britain's Frank Whittle. Wing span was 23 ft 3½ in (7·20 m), length 24 ft 6½ in (7·48 m), loaded weight 4396 lb (1998 kg) and maximum speed about 435 mph (700 km/hr).*

The Bell X-1 *(originally XS-1) research aircraft is often regarded as second in importance only to the Wright Flyer in the history of powered aeroplane flight, as it was the first to exceed the speed of sound (Mach 1) on 14 October 1947. The pilot was Capt Charles 'Chuck' Yeager, USAF, and the X-1 was named* Glamorous Glennis *after his wife. It was air-launched from under the belly of a B-29 bomber 'mother-plane', after which the 6000 lb (2722 kg) thrust rocket-engine consumed the aircraft's 8177 lb (3709 kg) of fuel in 2½ minutes. Wing span was 28 ft (8·54 m), length 31 ft (9·45 m) and loaded weight 13 400 lb (6078 kg). The X-1 later attained 967 mph (1556 km/hr).*

Mali   Current equipment of the Mali Republican Air Force includes about five MiG-17s; two Mi-4 helicopters; five to seven transports and some trainers. Strength: five combat aircraft, 400 personnel.

Mauritania   Current equipment of the Mauritanian Islamic Air Force includes three Britten-Norman Defenders; two Reims-Cessna FTR 337 Miliroles; and about ten transports. Strength: five armed aircraft, 150 personnel.

Mexico   After operations against rebel forces in 1911 by an American mercenary pilot, Hector Worden, the Mexican Government was encouraged to form a small air force which became the Mexican Aviation Corps in 1915, later to be enlarged into the Mexican Air Force (Fuerza Aerea Mexicana). Current equipment includes Lockheed T-33As; Beech AT-11 Kansan reconnaissance aircraft; 40 transports; over 100 trainers (some armed); and 40 helicopters. Strength: 15 combat aircraft, 6000 personnel. Navy operates five Convair Catalinas; four Grumman HU-16As; several helicopters and other aircraft.

Mozambique   Initial equipment of the air force of the FPLM includes seven Zlin 326 armed primary trainers; two Aérospatiale Alouette III helicopters and about ten Nord Noratlas and Douglas C-47 transports.

Mongolia   Current equipment includes several MiG-15s; 30 transports; ten Mil helicopters and several Yak-11 and -18 trainers.

Morocco   Following the emergence of Morocco as an independent State in 1956, the Royal Moroccan Air Force was established on 19 November that year, with a variety of light aircraft and a small number of personnel trained in France and Spain. Current equipment includes Dassault Mirage F1s; Northrop F-5As and RF-5As; Aérospatiale Magister armed trainers; about 30 transports; more than 60 trainers; up to 100 helicopters. About 12 MiG-17s and

Ilyushin Il-28s are in store. Strength: 70 combat aircraft, 5000 personnel.

Nepal   Current equipment includes two Aérospatiale helicopters and five transports.

Netherlands   Origins of the Royal Netherlands Air Force date back to the last century when, in 1886, the Dutch Army formed a balloon unit for artillery observation duties. Aircraft were first used experimentally during military manoeuvres in September 1911, and on 1 July 1913 an Aviation Division of the Royal Netherlands Army was established by Royal Warrant, to be based at Soesterberg. The Naval Aviation Arm (Marine Luchtvaartdienst) was formed on 18 August 1917. Current equipment of the Royal Netherlands Air Force includes about 90 Lockheed Starfighters; about 70 Canadair F-5s; about 90 Aérospatiale Alouette III and MBB BO 105 helicopters; transports; trainers and other aircraft. General Dynamics F-16As are on order. Strength 17 500 personnel. Navy operates about 14 Lockheed Neptunes; eight Breguet Atlantics; about 20 Agusta-Bell Iroquois, Westland Wasp and Sea Lynx helicopters; and a few other aircraft.

New Zealand   No formal military aviation corps existed in New Zealand during the First World War, pilots serving instead with the RFC and RNAS. Continuing efforts to pursue military aviation during the 1920s led to the formation of the New Zealand Permanent Air Force and Territorial Air Force in June 1923. The Royal New Zealand Air Force was constituted on 1 April 1937. Current equipment includes McDonnell Douglas Skyhawks; BAC Strikemasters; Lockheed Orions; 20 transports; about 25 helicopters; and trainers. Strength: 31 combat aircraft, 4250 personnel. Navy operates two Westland Wasps.

Nicaragua   In about 1923 the Nicaraguan Army was provided with a small number of Curtiss JN-4s and DH-4s by the USA, but little was done to form a regular air arm until 9 June 1938 when, under American guidance, the Nicaraguan Air Force (Fuerza Aerea de la Guardia Nacional) formally came into being. Current equipment includes Lockheed T-33s; North American T-28Ds; Douglas Invaders; nine transports; six helicopters and 15 other aircraft. Strength: 15 combat aircraft, 1500 personnel.

Nigeria   Following establishment of the Nigerian Federation in 1960, the Federal Nigerian Air Force was formed in 1964. Initial assistance from an Indian mission was followed by provi-

Canadair/Northrop NF-5s of the Royal Netherlands Air Force

sion of both aircraft and training personnel from West Germany. Current equipment includes MiG-21s and MiG-17s; a few Ilyushin Il-28s; 25 transports, including six Lockheed C-130H Hercules; 15 helicopters; 35 trainers. Strength: 20 combat aircraft, 5000 personnel.

Niger Republic Current equipment of the Niger Air Force (formed in 1960) includes about ten transports and a few light aircraft. Strength: 100 personnel.

Norway In mid-1912 a German Taube (named *Start*) was purchased by five Norwegian naval officers who presented it to the Royal Norwegian Navy. Almost simultaneously a Maurice Farman (named *Ganger Rolf*) was presented to the Royal Norwegian Army by Norwegians resident in France. Army and Navy flying schools were founded in 1914 and in the following year the Army Air Service (Haerens Flyvapen) and Naval Air Service (Marinens Flyvevaesen) were formed. Current equipment of the Royal Norwegian Air Force includes Northrop F-5As and RF-5As; Lockheed/Canadair Starfighters; Lockheed Orions; 36 helicopters; 12 transports; 35 trainers and other types. On order are General Dynamics F-16As. Strength: 105 combat aircraft, 9000 personnel.

Oman Following limited support provided by the RAF, at the request of the Sultan of Muscat and Oman, to crush the rebellion raised by Imam Ghalib, the Sultan of Oman's Air Force was established in 1958. Initial equipment comprised five Provost trainers and two Pioneer STOL utility aircraft. Current equipment includes SEPECAT Jaguars; BAC Strikemasters; Hawker Hunters; 28 transports; about 30 helicopters and a few other aircraft. Strength: 50 combat aircraft.

Pakistan The emergence of Pakistan as a dominion in July 1947 was accompanied by the formation of the Royal Pakistan Air Force. Two squadrons were established with former members of the Royal Indian Air Force. Current equipment of the Pakistan Air Force includes Chinese-built F-6s (MiG-19s); Dassault Mirage IIIs, III-Rs and 5s; North American F-86Fs and Canadair Sabres; Martin B-57Bs; Breguet Atlantics; Westland Sea Kings (ASW); nine transports; a few helicopters; 100 trainers. Strength: 230 combat aircraft, 17000 personnel. Army Aviation Wing has about 50 helicopters and 50 light aircraft.

Panama Republic The Panamanian Air Force was formed in January 1969 with American assistance. Principal tasks are to assist coast-

guard and police units, and initial equipment comprised Cessna U-17A lightplanes for liaison duties and two Douglas C-47 transports. Current equipment includes about 12 transports; 12 helicopters and smaller aircraft.

Papua New Guinea Initial equipment for the Air Force, provided by Australia in 1975, included four Douglas C-47 transports.

Paraguay The Paraguayan-Bolivian War of 1932 encouraged the Paraguayan Army to acquire some Potez 25 biplanes and, under the guidance of an Italian Air Mission, these were operated by mercenary pilots; by the end of the war in 1935 a regular Air Force, Fuerzas Aereas Nacionales, had been formed. Current equipment includes about 12 North American T-6 armed trainers; 20 transports; 17 helicopters; 16 trainers. Strength: 1800 personnel.

Peru Although financial appropriations were provided for the training of military pilots as early as 1912, it was not until late in 1919 that a military air corps was formed under the aegis of a French Air Mission and with 24 British and French aeroplanes. The Peruvian Naval Air Service followed in 1924, but on 20 May 1929 the two were combined to form the Cuerpo de Aeronautica del Peru. Current equipment of the Peruvian Air Force includes Hawker Hunters; North American F-86 Sabres; BAC Canberras; Cessna A-37Bs; Dassault Mirage 5s; Lockheed AT-33s; Grumman Albatross; 85 transports; about 40 helicopters; many trainers and a few other aircraft. Strength: 100 combat aircraft, 9000 personnel. Aircraft on order include Sukhoi Su-20s. Navy operates nine Grumman Trackers and about 40 other aircraft. Army operates 20 aircraft.

Philippines The Air Corps of the Philippine Army was formed on 2 May 1935 as a branch of the Philippine Constabulary. The Philippine Air Force came into being on 3 July 1947, exactly one year after the inauguration of the Philippine Republic. Current equipment includes Northrop F-5As; North American F-86F Sabres; about 60 helicopters; about 60 transports; 48 SIAI-Marchetti SF-260MP and SF-260WP Warrior attack/trainers; about 50 other trainers and several light aircraft. A few Grumman Albatross are used for search and rescue missions. Strength: 55 combat aircraft, 15000 personnel.

Poland A Polish squadron was incorporated in the Polish Army Corps in 1917, but a fully integrated Air Force was not formed until 29 September 1919 when, under Brigadier-General Macewicz, the new force operated against the USSR. Current equipment includes about ten regiments of MiG-21 interceptors; four regiments of Sukhoi Su-20 and Su-7B and MiG-17 ground attack aircraft; Ilyushin Il-28s and MiG-21s for ECM and reconnaissance; about 40 helicopters; 30 transports; and trainers. Strength: 800 combat aircraft, 55000 personnel. Naval air force has about 50 MiG-21s and MiG-17s, and 20 helicopters.

Portugal The Portuguese Air Force (Forca Aérea Portuguesa) owes its origins to funds publicly subscribed in 1912 which were used to purchase a small number of British and French aircraft, and a school was established at Villa Nova da Rainha. In 1917 army and naval air arms (Arma de Aeronáutica and Aviação Maritima) came into being. Current equipment includes North American F-86F Sabres; Fiat G 91Rs; Lockheed Neptunes; Cessna Skymaster counter-insurgency aircraft; 26 transports; 48 helicopters; 75 trainers. Strength: 88 combat aircraft, 12000 personnel.

Qatar Following withdrawal of UK military forces from the Persian Gulf area at the end of 1971, Qatar elected to build up independent defence forces with British assistance. The inauguration of the Qatar Public Security Forces Air Arm may be dated from the receipt of its first two Westland Whirlwind 3 helicopters in March 1968. Current equipment includes Hawker Hunters; nine Westland helicopters; and one Britten-Norman Islander. Strength: three combat aircraft.

Rhodesia In 1936 Southern Rhodesia organised the basis of an Air Section of the Permanent Staff Corps at Salisbury. This administered substantial contributions to the RAF during the Second World War. The Southern Rhodesian Air Force changed its title to the Royal Rhodesian Air Force in October 1954. Following the dissolution of Federation with Nyasaland in March 1963, the title of this force changed again to the Rhodesian Air Force. Current equipment includes Hawker Hunters; de Havilland Vampires; BAC Canberras; 12 helicopters; 15 transports; and about 30 trainers (BAC Provosts) used also for reconnaissance. Strength: 32 combat aircraft, 1200 personnel.

Romania Romania was one of the first countries in the world to form a regular air force using aeroplanes, its army having established a Flying Corps in late 1910. Current equipment includes 100 MiG-21s; 100 MiG-17s; 50 Sukhoi Su-7Bs; 33 transports; about 170 helicopters; many trainers and other aircraft. Strength: 24000 personnel.

Rwanda Republic Current equipment includes three armed Aérospatiale Magisters; three transports; two helicopters and three liaison aircraft. Strength: 150 personnel.

Saudi Arabia The establishment of a small air force was proposed by Ibn Saud in 1923 and this was equipped by Britain with a small number of DH9s under an agreement for collaboration which terminated in 1933. The Royal Saudi Air Force dates from 1950 when a British Mission reorganised the force and supplied a nucleus of light transport and training aircraft. Current equipment includes 70 Northrop F-5E Tiger IIs; 34 English Electric Lightnings; 30 BAC Strikemasters; 450 helicopters (most of which are Bell HueyCobras); 30 transports; four Lockheed Hercules tankers and many trainers. Strength: 10000 personnel.

Senegal A former French colony, Senegal became independent in November 1958. The Senegal Air Force may be said to date from late 1960 when some C-47s, Broussards and light helicopters were received from France. Current equipment includes six transports; five helicopters and five liaison aircraft. Strength: 200 personnel.

Sierra Leone Current equipment includes four SAAB-Scania MFI-15s and three Hughes helicopters.

Singapore Following the British Government's decision to withdraw the majority of its armed forces from the Far East before the end of 1971, Singapore began to build up strong military forces. Recruiting for the Singapore Air Defence Command began in April 1968. Current equipment includes Hawker Hunters; McDonnell Douglas Skyhawks; BAC Strikemasters; 25 trainers; six Short Skyvan transports and seven Aérospatiale Alouette III helicopters. Strength: 94 combat aircraft, 3000 personnel.

Somali Republic Current equipment of the Air Force, built up since 1963 with Soviet assistance, includes Ilyushin Il-28s; MiG-21s, MiG-17s and MiG-15s; 12 transports; Mil helicopters and about 30 trainers. Strength: 30 combat aircraft, 2500 personnel.

South Africa After six officers of the Union Defence Forces, who had received flying training at a flying school at Kimberley, had been sent to join the RFC on the outbreak of the First World War, the South African Aviation Corps was formed under Major Van der Spuy in 1915. Current equipment of the South African Air Force includes Dassault Mirage III-Cs, III-Es and III-Rs; Dassault Mirage F1s; Canadair Sabres; Hawker Siddeley Buccaneers; Avro

Shackletons; BAC Canberras; Atlas Impalas (Aermacchi MB326s); 30 transports; 120 helicopters; many trainers and other types. Strength: 175 combat aircraft, 8500 personnel.

South Yemen, Republic of The Air Force of the Southern Yemen People's Republic dates from the assumption of power by the National Liberation Front in Aden, and British withdrawal on 29 November 1967. Initial equipment consisted of four Douglas C-47s converted from ex-airline DC-3s. Current equipment includes 15 MiG-17s; 12 MiG-21s; ten transports; several helicopters and other aircraft. Strength: 2000 personnel.

Spain Spain formed its first military aviation force in 1896 with the establishment of the Servicio Militar de Aerostación, a captive balloon section. Later, in 1910, plans were laid for the formation of the Aeronáutica Militar Española and in March the following year four French aeroplanes formed the new air force's initial equipment. Current equipment of the Spanish Air Force includes Dassault Mirage IIIs; McDonnell Douglas Phantom IIs; Dassault Mirage F1s; Northrop SF-5As/SRF-5As; Hispano Saetas; Lockheed Orions; Grumman Albatross; about 130 transports; Canadair CL-215s for search and rescue; about 30 helicopters; many trainers and other aircraft. Strength: 200 combat aircraft, 35000 personnel. Navy operates a few Hawker Siddeley Matadors (Harriers) and about 80 helicopters. Army has about 60 helicopters.

Sri Lanka The original Royal Ceylon Air Force was formed on 10 October 1950 with de Havilland Vampire trainers. Current equipment

includes five MiG-17s; 12 transports; about 15 helicopters; and about 17 other aircraft. Strength: 2000 personnel.

Sudan After the proclamation confirming Sudan as a Republic on 1 January 1956 steps were taken to form an air force. Four light aircraft were presented by Egypt in the following year with the founding of the Sudanese Air Force. Current equipment, supplied mainly from the Soviet Union and China, includes MiG-21s and MiG-17s; BAC Strikemasters; about 18 transports; 14 Mil helicopters; and other types. Strength: 33 combat aircraft, 2000 personnel.

Sweden Although the Flygvapnet of today came into being on 1 July 1926, military and naval aviation in Sweden started in 1911 with the presentation of single military and naval aircraft to the nation. Flying schools were formed at Axvall and Oscar Fredriksborg. In 1914 a military flying echelon, the Fälttelegrafkarens Flygkompani ('Field Telegraph Aviation Company') was formed. Current equipment includes about 325 SAAB J35 Drakens and 50 S35 Drakens; 130 SAAB AJ37 Viggens and 15 SH37 Viggens; 40 SAAB A32 Lansens and 30 S32 Lansens; 20 SAAB Sk60B/C reconnaissance/attack aircraft and 120 Sk60A trainers adaptable for attack; 13 transports; 16 helicopters and other types. Strength: 8200 personnel. The Navy has 37 helicopters. The Army has 39 helicopters and 35 light aircraft.

Switzerland The Swiss Fliegertruppe was established on 31 July 1914 at Buedenfeld. Initial equipment comprised one indigenous and three

SAAB AJ37 Viggen attack aircraft of the Swedish Air Force

French monoplanes. Current equipment includes 140 de Havilland Venoms; 125 Hawker Hunters; over 50 Dassault Mirage IIIs; nearly 100 Aérospatiale Alouette II and III helicopters; about 30 transports; about 170 trainers; and other aircraft. On order are 72 Northrop F-5E Tiger IIs and F-5Fs. Strength: 3000 regular personnel.

Syria   Current equipment includes MiG-23s, MiG-21s and MiG-17s; Ilyushin Il-28s; Sukhoi Su-7s; Kamov Ka-25 ASW helicopters; about 20 transports; 50 helicopters; and trainers. Strength: 425 combat aircraft, 25 000 personnel.

Tanzania   The Tanzanian People's Defence Force Air Wing had its origin in a small military air arm built up with, successively, West German and Canadian assistance from 1964. Major expansion followed, with Chinese aid. Current equipment includes Chinese-built F-8s (MiG-21s), F-6s (MiG-19s) and F-4s (MiG-17s); 11 transports; four helicopters and 12 trainers. Strength: 33 combat aircraft, 1000 personnel.

Thailand   The Royal Siamese Flying Corps was formed on 23 March 1914 after three officers of the Royal Siamese Engineers, who had received their flying training in France, returned home. Eight French aircraft provided its initial equipment. Current equipment of the Royal Thai Air Force includes Northrop F-5s; Cessna A-37Bs; North American T-28Ds; Rockwell OV-10C Broncos; Fairchild AU-23 Peacemakers; North American T-6Gs; Lockheed RT-33As; over 100 helicopters (mostly Bell Iroquois); over 50 transports; many trainers and other aircraft. Strength: 175 combat aircraft, 42 000 personnel. Navy operates 12 Grumman Albatross and Trackers. Army operates about 120 helicopters (mostly Iroquois) and many Cessna Bird Dogs. Police also use 18 aircraft and over 25 helicopters.

Togo   The Force Aerienne Togolaise was created with French help after independence was achieved in 1960. Current equipment includes three Embraer EMB-326 Xavante armed trainers; six transports; three helicopters and five trainers.

Tunisia   The Tunisian Air Force was established following the attainment of independence in 1956, and 15 SAAB-91D Safir aircraft for ab initio training were ordered mid-1960. Current equipment includes Northrop F-5E Tiger IIs; Aermacchi MB326Ks; SIAI-Marchetti SF260W Warrior strike/trainers; three transports; ten helicopters; 15 trainers. Strength: 28 combat aircraft, 2000 personnel.

Turkey   Foreign pilots flew the small number of aeroplanes operated during the Balkan Wars of 1912–13. The Turkish Flying Corps came

McDonnell Douglas A-4 Skyhawks and F-4 Phantoms of the US Marine Corps, operating from the carrier USS *John F. Kennedy*

Hawker Siddeley AV-8A Harrier of the US Marine Corps (Norman E Taylor)

into formal being early in 1915, almost exclusively manned by German crews in German aircraft. Current equipment of the Turkish Air Force includes Northrop F-5As; Lockheed F-104G/S Starfighters; McDonnell Douglas Phantom IIs; Convair F-102A Delta Daggers; North American F-100 Super Sabres; about 50 transports; 23 helicopters; and many trainers. Strength: 400 combat aircraft, 45 000 personnel. Navy operates about 12 Grumman Trackers and three Agusta-Bell 205AS helicopters. Army operates over 200 helicopters and many fixed-wing aircraft.

Uganda Following attainment of independence in 1962, it was decided to form an air force to support the Uganda Rifles. Initial finance was provided in 1964, for formation of the Uganda Army Air Force and for expansion of the Police Air Wing. Current equipment includes a few MiG fighters and other types. Some of the ten transports and 18 helicopters are operated by the Police Air Wing. Strength: 600 personnel.

United States of America Balloons were used by both sides in the American Civil War between 1861 and 1863. An Aeronautical Division of the Signal Corps (with a personnel strength of three) was formed in August 1907, and the first aeroplane, a Wright biplane, was accepted by the Army on 2 August 1909. A Signal Corps Aviation Section was authorised on 18 July 1914 and, following outstanding pioneer maritime flying, a Naval Office of Aeronautics was established on 1 July 1914. The Department of the Air Force was activated within the Department of Defense in September 1947, coequal with the Army and Navy. It is responsible for the strategic deterrent force of

1054 Minuteman and Titan intercontinental ballistic missiles. Current aircraft in service include a total of 421 Boeing B-52 and General Dynamics FB-111 strategic bombers; 315 Convair F-106 Delta Dart home defence interceptors; nearly 6000 tactical fighter/attack and reconnaissance aircraft, comprising General Dynamics F-111s, Vought A-7 Corsair IIs, Fairchild A-10s, McDonnell Douglas F-4 Phantoms and F-15 Eagles, and Republic F-105 Thunderchiefs; McDonnell Douglas RF-4 Phantoms; F-4G Wild Weasel ECM aircraft; Boeing E-3A airborne warning and control aircraft; Boeing E-4 airborne command posts; about 30 Lockheed SR-71A and U-2 strategic reconnaissance aircraft; 370 Cessna A-37B and North American OV-10 Bronco counter-insurgency aircraft; about 1200 Lockheed C-5 Galaxy, C-130 Hercules and C-141 StarLifter transports and other types; several hundred helicopters; more than 2000 trainers, and other types. Strength: total of 9239 aircraft, 820 000 personnel. The US Navy and Marine Corps have 4930 operational aircraft, including 1135 attack, 623 fighter, 112 ASW, 369 patrol, 106 warning and 213 transport aircraft, and 1087 helicopters. The Army has between 6700 and 9000 aircraft, mainly helicopters.

Upper Volta Current equipment includes five transports and five other aircraft.

Uruguay A Department of Military Aviation was created on 20 November 1916, as well as a School of Military Aeronautics at San Fernando. Current equipment includes Lockheed F-80C Shooting Stars; Lockheed AT-33s; 25 transports; four helicopters; 18 trainers and 19 liaison aircraft. Strength: 12 combat aircraft, 2000 personnel. Navy operates three Grumman Trackers; six helicopters, seven trainers and three other aircraft.

USSR The Imperial Russian Flying Corps was formed in 1910, together with an Army Central Flying School at Gatchina and a Naval Flying School at Sevastopol. This Flying Corps disintegrated soon after the outbreak of the Revolution in November 1917. In two orders issued on 20 November 1917 and 24 May 1918 the Chief Administration of the Workers' and Peasants' Red Air Fleet (Glavnoe Upravlenie Raboche-Krestyanskogo Krasnogo Vozdushnogo Flota, GUR–KKVF) was formed. Today there are three independent major components. The Strategic Nuclear Forces include 1527 ICBMs; 600 MRBMs and IRBMs; 35 Myasishchev M-4, 500 Tupolev Tu-16, 170 Tu-22, 110 Tu-95 and 50 supersonic Tupolev 'Backfire' bombers; and 70 M-4 and Tu-16 tankers. The Air Defence Force has about 2600 interceptors, including MiG-25s, MiG-23s, MiG-17s, Sukhoi Su-15s, Su-11s and Su-9s, Tupolev Tu-28Ps and Yakovlev Yak-28Ps; and 12000 surface-to-air missiles. The Air Force has about 4500 tactical combat aircraft, including Ilyushin Il-28s, MiG-27s, MiG-25s, MiG-23s, MiG-21s, MiG-17s, Sukhoi Su-19s, Su-17s, Su-7s and Yakovlev Yak-28s; about 3200 Mil helicopters, including Mi-24 gunships and assault helicopters; about 1500 transports, half of them Antonov An-12s; and many training and other types. Air Force has 450000 personnel; Air Defence Force 550000 personnel. The Soviet Naval Air Arm has about 640 combat aircraft, including 475 maritime patrol and attack bombers; 165 ASW aircraft and amphibians; plus about 250 Kamov Ka-25 ASW helicopters and 200 transports. The Yakovlev Yak-36 VTOL carrier-based attack/reconnaissance aircraft is entering service.

Union of Arab Emirates The UAE has a small defence force made up of elements from Abu Dhabi, Dubai, Qatar, Ras Al Khaimah and Sharjah. Current equipment includes three Bell JetRanger and four Bell 205 helicopters.

Venezuela The Venezuelan Military Air Service was established on 17 April 1920 and a flying school was set up at Maracay; the first pilots underwent training in 1921 under French supervision. Current equipment includes Canadair CF-5A/Bs; North American F-86K Sabres; Dassault Mirage IIIs and 5s; BAC Canberras; Rockwell Broncos; up to 60 transports; about 37 helicopters and over 55 trainers. Strength: 90 combat aircraft, 6000 personnel. Navy operates four Grumman Albatross; three Grumman

Trackers; and a few other aircraft. Army operates about 20 helicopters.

Vietnam The Vietnamese Air Force owes its origins to the setting up under French supervision of flying training facilities at the Armée de l'Air base at Nha Trang in 1951. Observation squadrons were brought into being and these participated with the French against Vietminh forces between 1952 and 1955. The current status of the very large number of US aircraft supplied to the former South Vietnam forces is unknown. Aircraft believed now to be operational in the unified state include Ilyushin Il-28 bombers; MiG-21, MiG-19 and MiG-17 fighters; Sukhoi Su-7 attack aircraft; about 40 transports; 25 helicopters; and 30 trainers. Strength: 190 combat aircraft, 12000 personnel.

Yemen Arab Republic (North) Current equipment includes 12 MiG-17s; 12–15 Ilyushin Il-28s; several transports and helicopters; and about 15 Yak-11 trainers. Strength: 1500 personnel.

Yugoslavia Although Yugoslavia as a sovereign State only emerged in 1918, its aviation origins lay in the Serbian Military Air Service formed in 1913 with the return home of six Serbian army officers from France where they had received their flying training. Current equipment of the Yugoslav Air Force includes 100 MiG-21s; up to 125 SOKO Jastreb light attack aircraft; 50 North American F-86D/K Sabres; 20 Lockheed RT-33As; about 180 helicopters; 40 transports; many trainers and other types. Strength: 30000 personnel.

Zaire Republic The Congolese Air Force was formed originally in mid-1961, then equipped with a small number of light aircraft. It has expanded considerably since uniting with the Air Force of the former Central Government of Katanga. Current equipment includes 14 Dassault Mirage 5s; 13 Aermacchi MB-326 armed trainers; 21 transports, including five Lockheed C-130 Hercules; 18 helicopters; and trainers. Strength: 1500 personnel.

Zambia The Zambian Air Force originates from defence elements allocated to Northern Rhodesia after its seccession from the Central African Federation in 1963. Initial equipment comprised four Douglas C-47s and two BAC Pembrokes transferred from the Royal Rhodesian Air Force. Current equipment includes six SOKO Jastrebs and Galebs, and 18 Aermacchi MB326GB armed trainers; about 40 transports; 45 helicopters; and eight trainers. Strength: 1500 personnel.

SNECMA Coléoptère, a VTOL jet prototype with a ring-wing, first flown in May 1959

SECTION 10

POST WAR - RESEARCH

As the early sections of this book have shown, research and experiments in flying have been under way since the first real-life Daedalus strapped on a pair of home-made wings and jumped from a high place into oblivion. Progress began with Cayley, Lilienthal, the Wrights, Blériot and the other pioneers of our own century; and the period since the Second World War has brought advance and change beyond even their dreams. There are many reasons for this. Not least has been the spur of the so-called arms race.

For a time after the First World War, there was still talk of the 'war to end wars'; but it did no such thing. Rather, the manner of its victories, and the conditions imposed by the Peace agreements which followed, made a Second World War inevitable. In its turn, the end of the Second World War might have brought only a decade or two of growing tension before a third world conflict, had not America demonstrated in the most terrible manner the effects of atomic warfare.

So began the 'cold war', followed by the deterrent policy of 'peace through fear'. This depends on keeping potential enemies in such fear of mutual annihilation in a nuclear exchange that they will never see any advantage to be gained from aggression. Unfortunately, in an age of immense techno-logical advances in aviation and rocketry, the vital balance of power can be maintained only at crippling expense. Each new product of military research necessitates still greater expenditure to utilise or counter it. So we have the two modern super-powers, America and Russia, which alone have the economic resources to capitalise on inventive capability.

In the early post-war years, when inventiveness and engineering skill held the key to progress, Britain set the pace in many ways. The United States was first to prove that the 'sound barrier' could be penetrated by the brute force of rocket power allied to human courage. It was the Fairey Delta 2 which combined British leadership in jet propulsion with design genius to show that passing the speed of sound, Mach 1, need result in nothing more alarming than the flicker of a needle or two on the instrument panel. Money began to run out by the time the Fairey Rotodyne pioneered the whole new field of vertical take-off and landing (VTOL) for commercial operations. It dried up completely before the British Aircraft Corporation's TSR.2 could keep Britain in the military 'big league' with its supersonic bombing capabilities.

The United States, on the other hand, launched into new generations of aviation research which investigated every kind of technique for achieving VTOL, without producing anything better than the relatively simple helicopter; and poured countless millions of dollars into a giant Mach 3 bomber that remained but a prototype. In a last fling at more reasonable economic levels, Britain produced the first thoroughly practical VTOL combat aircraft in the Harrier, which America decided it had to buy. With its French neighbours, it then evolved the Concorde supersonic airliner, unbeaten as a technological triumph but not large enough to be a success economically, as neither nation could afford to design and build it to carry 250 passengers rather than 120.

Meanwhile, the United States has progressed relentlessly from the Mach 1·015 of 'Chuck' Yeager's little Bell XS-1 of 1947 to Mach 6·72 with the North American X-15A-2, and is now preparing to launch the remarkable Space Shuttle Orbiter into a routine research lifetime at Mach 25 or more. Already there are aircraft designers who see little point in building airliners that will fly at speeds between the Concorde's Mach 2 and the Mach 6 that is entirely practicable if proven engineering technology is combined with liquid hydrogen fuel. Whether even the United States or the Soviet Union could afford to produce such aircraft, which would bring Australia within three or four hours of Western Europe – or would know how to utilise the immense work capacity that it offers – is something that only the future will reveal.

Before listing some of the more significant post-war research aircraft, it is worthwhile recalling a few of their predecessors which paved the way for them to follow:

The first rocket-powered aeroplane in the world was the sailplane *Ente* ('Duck'), powered by two Sander slow-burning rockets and built by the Rhön-Rossitten Gesellschaft of Germany. Piloted by Friedrich Stamer, it made a flight of just over 0·75 mile (1·2 km) near the Wasserkuppe Mountain in about 1 min on 11 June 1928. The rocket-powered glider flown by Fritz von Opel at Rebstock, near Frankfurt, is often stated as being the world's first rocket aeroplane, but did not fly until 30 September 1929. It then covered a distance of about 2000 yd (1830 m), attaining a speed of 100 mph (160 km/h) for a short period.

The first successful liquid-fuel rocket aircraft in the world was the German DFS 194 which, having been conceived by the Sailplane Research Institute in 1938 under Professor Alexander Lippisch, was taken over by Messerschmitt AG at Augsburg and flown in 1940 by Heini Dittmar. It was powered by a 600 lb (272 kg) thrust Walter rocket.

The first American rocket-powered military aircraft was the Northrop MX-324, which was first flown under rocket power by Harry Crosby on 5 July 1944. It was powered by an Aerojet XCAL-200 motor fuelled by mono-ethylaniline. It had flown originally as a glider, piloted by John Myers, on 2 October 1943.

The first research aircraft to be designed for flying at 1000 mph (1600 km/h) was the Miles M-52. Development began in 1943, and by February 1946 the detail design was virtually completed. Construction was well underway when the project was cancelled due to economic problems and the belief by some officials that it should have been designed with swept instead of very thin bi-convex straight wings. However, models of the M-52 flown in 1947–8 showed that the aircraft could have achieved its aim. Power for the M-52 was to have been provided by one Power Jets W2/700 turbojet engine, with augmentor and afterburner, developing up to 4100 lb (1860 kg) thrust.

The first jet-powered aircraft and the first British aircraft to exceed Mach 1 was the de Havilland DH108, three examples of which were built to investigate the stability and control problems of swept wings, so providing information for the design of the de Havilland Comet 1 airliner. The first DH108 made its maiden flight on 15 May 1946, using a standard de Havilland Vampire fighter fuselage and powered by one 3750 lb (1700 kg) thrust Goblin 4 turbojet engine. This first aircraft was used to provide data on the slow-flying characteristics of the swept wings; the second and third were used for high-speed flying. The last aircraft recorded a speed of Mach 1·0–1·1 on 6 September 1948 while in a dive from 40 000 ft (12 200 m).

The first French aircraft to be designed specifically for stratospheric research flying was the Aérocentre Belphégor, which flew for the first time on 6 June 1946. Powered by a single Daimler-Benz DB610 engine, developing 3000 hp, it had pressurized accommodation for five persons in its bulbous fuselage, including two research members; the pilot was situated in a cupola above the main cabin. With an all-up weight of 22 050 lb (10 000 kg), the Belphégor could fly to an altitude of 42 000 ft (12 800 m).

The most unusual bomber programme initiated after the Second World War involved the series of Northrop flying wings. The first full-size prototype XB-35 flew on 25 June 1946. Designed as a long-range bomber, it comprised a cantilever aluminium sweptback wing, constructed in one piece. Directional control was via drag-inducing double-split flaps at the wingtips, and elevons and leading-edge fixed wingtip slots were fitted. The crew of seven were situated in a centre-section nacelle; six electrically-operated gun turrets (four remotely-controlled at outer wing stations) provided defensive armament. Powered by four turbo-supercharged Pratt and Whitney Wasp-Major piston-engines of 3000 hp each, the XB-35 had an all-up weight of 209 000 lb (94 800 kg) and a wing span of 172 ft (52·43 m). Its design had been started in 1942, followed by testing of four twin-engined scale models designated N-9M. Following the first flight of the prototype XB-35, 14 development aircraft were ordered by the USAF as YB-35s; two of these were converted into YB-49s, each with eight jet engines, and one into the YRB-49A with six jet engines.

Northrop YRB-49A flying-wing bomber

Although 30 RB-49s were ordered subsequently for operational service with the USAF, these were cancelled in 1949.

The first piloted aircraft to fly faster than the speed of sound was the Bell X-1 (initially XS-1) on 14 October 1947. Powered by one Reaction Motors E6000–C4 rocket motor, the X-1 achieved a speed of 670 mph (1078 km/h) on this occasion (equivalent to Mach 1·015 at 42 000 ft (12 800 m)), piloted by Captain Charles 'Chuck' Yeager, after being air-launched from a B-29 Superfortress bomber. Altogether 231 flights were made by X-1 series aircraft, the first on 9 December 1946, and Mach 2·435 was reached on 12 December 1953 by the X-1A, which also flew to an altitude of 90 000 ft (27 430 m) in June 1954. The last version of the X-1 series was the X-1E.

The first jet-powered flying-boat in the world was the British Saunders-Roe SRA/1. See Post Second World War Military Aviation.

Designed to obtain air-load measurements, the Douglas Skystreak raised the World Speed Record by 17 mph (27 km/h) to 640·60 mph (1030·95 km/h) on 20 August 1947, then to 650·78 mph (1047·33 km/h) on 25 August. The first of three Skystreaks flew initially on 28 May 1947, and automatic pressure-recording was carried out via 400 measurement points on the fuselage, straight laminar-flow wings and tail unit. In addition strain gauges were attached to the wings and tail unit.

First post-war convertiplane experiments carried out in Britain were performed by the five-seat Fairey Gyrodyne, which made its first free flight on 7 December 1947. Both the main rotor and variable-pitch propeller (which was mounted at the tip of the starboard stub-wing) were driven by a single Alvis Leonides piston-engine of 525 hp. The stub-wing propeller arrangement enabled the Gyrodyne to be flown as an autogyro; on 28 June 1948 it raised the international speed record for helicopters to 124·5 mph (200 km/h). On 17 April 1949 the Gyrodyne crashed, due to rotor head fatigue, and the crew was killed. However, experiments continued with a second aircraft, called

Bell X-1 under its B-29 mother-plane

Douglas Skystreak

McDonnell Goblin parasite fighter

the Jet Gyrodyne, which was modified later to have a tip-jet rotor system and pusher propellers on the stub wings.

The first piloted aircraft to be flown at twice the speed of sound was the Douglas Skyrocket, which flew for the first time on 4 February 1948. Designed to investigate swept-back wings, it was powered by one Reaction Motors XLR-8 rocket motor of 6000 lb (2720 kg) thrust and one Westinghouse J34 turbojet engine of 3000 lb (1360 kg) thrust. The wings were of conventional subsonic configuration with a 35° sweepback. Altogether three Sky-rockets were built, and on 31 August 1953 one reached an altitude of 83 235 ft (25 370 m). How-ever, the most memorable flight of a Skyrocket was made on 20 November of the same year when the aircraft attained a speed of Mach 2·005 after being launched from a Boeing 'mother-plane' at 32 000 ft (9750 m).

The first jet-powered fighter to be specifi-cally designed as a parasite aircraft was the McDonnell XF-85 Goblin, which made its first free flight on 23 August 1948. Designed to be carried inside the forward bomb-bay of the Consolidated Vultee B-36 long-range bomber, the Goblin had an extremely short and stubby fuselage, with swept short-span wings and mul-tiple tail surfaces. It was launched and picked up via a retractable 'skyhook' on the Goblin which hooked on to a retractable trapeze on the bom-ber. Following one abortive attempt to hook back onto the B-36, which nearly ended in disaster, the first successful hook-on was made on 14 October 1948. Although several more successful hook-ons were achieved, and the Goblin had a speed of 520 mph (837 km/h), the project was cancelled after only two Goblins had been built. One of them remains on view at the

USAF Museum at Dayton, Ohio. Length of the Goblin was only 14 ft 10½ in (4·53 m); power was provided by a Westinghouse J34 engine of 3000 lb (1360 kg) thrust.

The Northrop X-4 Bantam was designed to investigate the subsonic flying characteristics of aircraft with swept wings but without a tail-plane. Two X-4s were produced, the first flying on 15 December 1948. The programme was successfully completed in April 1954, after some 60 flights had been made.

Built as an experimental ramjet-powered aircraft, the French Leduc 0.10 made its first powered flight on 21 April 1949 after being released over Toulouse from a Languedoc mother-plane. This carried the 0.10 above its fuselage on special strut mounts, producing the stream of air flowing into the engine which was necessary for the ramjet to work. On this occasion the 0.10 reached a speed of 422 mph (680 km/h) on 50 per cent power. Maximum speed achieved during a later flight was Mach 0·84. Three 0.10s were produced. Each had a tubular double-skinned fuselage, the outer shell forming the annular ramjet duct and the inner shell accom-modating the cockpit for the crew of two.

The first British delta-wing research air-craft was the Avro 707, which made its first flight on 4 September 1949. Designed to gain data on the flight characteristics of delta wings at low speeds, the Type 707 was basically a scale model of the then-projected Vulcan bomber. Following its destruction in an accident, the Type 707B was produced to continue low-speed research, first flying in September 1950. Two Type 707As then followed for research into high-speed flight; and the series was completed by a single Type 707C, a two-seat version built to give pilots training in flying delta-wing air-craft. The 707C first flew in mid-1953, powered by a Rolls-Royce Derwent engine of 3600 lb (1635 kg) thrust.

The Boulton Paul PIII and PI20 were small delta-wing aircraft built to investigate the delta wing at transonic speeds. The basic differences between the two aircraft were that the PI20 had an all-moving tailplane while the PIII had four airbrakes mounted on the fuselage and a nose probe. First flight of the PIII was on 10 October 1950.

The Fairey FDI was a small delta-wing research aircraft that was conceived originally as a ramp-launched vertical take-off fighter. Powered by a single Rolls-Royce Derwent turbojet engine of 3600 lb (1635 kg) thrust, it made its first flight on 12 March 1951 at Boscombe Down. When the VTO project was abandoned, the FDI was given a conventional undercarriage and was used for research until it was grounded after a landing accident.

The first jet aircraft flown with variable-sweepback wings was the Bell X-5, which flew for the first time on 20 June 1951. Design of the X-5 began in 1948, and two were built, the first crashing on 13 October 1953. The X-5 was powered by a single Allison J35-A-17 turbojet engine. The wing span varied from 18 ft 7 in (5·66 m) to 30 ft 10 in (9·39 m) in swept and unswept positions respectively.

Built to test the flight characteristics of swept wings at low speed, the Short SB5 produced data that was used in the design of the English Electric PI prototype and, subsequently, the production Lightning fighter. The wings on the SB5 were tested at four different angles of sweepback, ranging from 50° to 69°, with combinations of a high or low mounted tailplane with various angles of incidence. After making its maiden flight on 2 December 1952, the aircraft was still flying experimentally in the early 1960s.

Designed as a mixed-power research aircraft, to provide data for future interceptors of similar concept, the French Sud-Ouest SO 9000 Trident first flew on 2 March 1953, powered initially by two wingtip-mounted Turboméca Marboré II turbojet engines. These were later replaced by Dassault Viper turbojet engines of nearly double the power, and a SEPR 481 rocket motor of 9920 lb (4500 kg) thrust was installed subsequently in the rear fuselage. Testing of the rocket power unit began in April 1955, and the Trident eventually reached a speed of 1055 mph (1700 km/h).

The first aircraft to test the practicability of the aero-isoclinic wing was the Short SB4 Sherpa, which made its maiden flight on 4 October 1953. The wing was designed as a partially flexible structure, with all-moving tips which were used as both elevators and ailerons. Flight testing showed that the handling characteristics of the aircraft were very satisfactory.

The Convair Sea Dart was an experimental seaplane fighter fitted with delta wings and hydroskis. It set several 'firsts', being the first delta-winged seaplane and the first seaplane to exceed the speed of sound. The original aircraft was designated XF2Y-1 and made its first flight on 9 April 1953. It was joined subsequently by the development version, designated YF2Y-1, which exceeded Mach 1 on 3 August 1954 in a dive, shortly before being destroyed in an accident. Whereas the XF2Y-1 was powered by two Westinghouse J34 turbojet engines, each of 3400 lb (1542 kg) thrust, the later aircraft had two J46 engines of 6000 lb (2720 kg) thrust each, with afterburning. The YF2Y-1 proved sufficiently promising for the US Navy to order three more similar aircraft, but the whole concept of a seaplane fighter was soon abandoned.

Two experimental VTOL fighters built in America, and first flown in 1954, were the Lockheed XFV-1 and Convair XFY-1 'tailsitters'. Designed to the same specification, both had a maximum speed of about 500 mph (805 km/h) and were powered by Allison YT40-A engines driving co-axial contra-rotating propellers. The Lockheed XFV-1 was the first to fly, in March 1954; it had straight wings and a cruciform tail unit, which doubled as its landing gear as a sprung castoring wheel was fitted to the tip of each tail surface. The Convair XFY-1 made its maiden flight on 2 August 1954, and differed from its competitor mainly in having delta wings, with large delta tail surfaces above and below the fuselage forming a cruciform with the wings. Castoring wheels were positioned at the tips of these wing and tail surfaces, to form the vertical undercarriage. The pilots of both aircraft sat on gimbaled seats, enabling them to be in a near-upright position irrespective of whether the aircraft was in a horizontal or vertical attitude. The XFV-1 made only conventional take-offs; but many VTOL flights were made by the XFY-1, including transitions from vertical to horizontal flight, before the concept of a tail-sitting VTOL fighter was abandoned.

Lockheed XFV-1 'tail-sitting' experimental fighter

The first aircraft to set a world speed record of over 1000 mph (1600 km/h) was the Fairey Delta 2, the first example of which made its maiden flight on 6 October 1954. Built originally to investigate the problems encountered during transition from subsonic to supersonic speeds, each Delta 2 was powered by a Rolls-Royce Avon turbojet engine. The first aircraft had an Avon RA5 which gave 12000 lb (5445 kg) thrust, the second an RA28 which developed 1000 lb (450 kg) more thrust. The maximum speed attained by a Delta 2 was 1147 mph (1846 km/h). Subsequently, the original example was converted into the BAC 221, with new delta wings and control surfaces, new landing gear, longer fuselage and a hydraulically-actuated drooping nose. Data gained with this aircraft were used in the development of the Concorde airliner.

The first aircraft to fly at over Mach 3 in level flight was the Bell X-2, two examples of which were built for continued research at transonic and supersonic speeds. Built with a K-monel metal fuselage, and stainless-steel swept wings and tail unit, the first X-2 was destroyed after an explosion in its B-50 mother-plane, which was to air-launch the aircraft, resulted in the research aircraft having to be jettisoned. The second X-2 made its maiden flight on 18 November 1955, but was also destroyed, after several successful flights, on 27 September 1956, at the end of a test in which it recorded Mach 3·2.

Convair Sea Dart fighter taking off on its hydro-ski

The first tilt-rotor convertiplane to be flown was the Bell XV-3. Two examples were built, the first making its initial vertical flight on 23 August 1955. Powered by a single Pratt and Whitney R-985 engine of 450 hp, the XV-3 had a large fuselage that could accommodate four persons, and was fitted with fixed wings of 31 ft 3½ in (9·54 m) span. The rotor/propellers were mounted at each wingtip and could be directed upward or forward for vertical or horizontal flight respectively, via small electric motors. The first transition from vertical to horizontal flight was made on 18 December 1958; by 1966 more than 250 flights had been logged.

Built to test a turbojet-ramjet engine, which was designed to form the bulk of the aircraft's airframe, the French Nord 1500-02 Griffon II was a direct development of the earlier Griffon I, which had been powered by a conventional turbojet only. The Griffon II had a turbojet mounted inside the ramjet, to propel it to the speed at which the ramjet could ignite and provide power. It flew for the first time on 23 January 1957, and exceeded Mach 1 on 17 May, with its ramjet power on. Over 200 flights were made by the aircraft, culminating in a flight of Mach 2·19 on 13 October 1959, at which speed the ramjet engine developed four-fifths of the aircraft's total thrust.

The Douglas X-3 was built to investigate the effectiveness of turbojet engines and short-span double-wedge wing and tail surfaces at very high altitudes, and to study thermodynamic heating at speeds up to Mach 3. Construction of the aircraft caused many problems, as it had to be built mostly of then-novel titanium. To measure the pressure on the airframe during flight, a huge number of pin-hole orifices were positioned strategically over the airframe, and temperature and stress were also measured at many locations. Powered by two Westinghouse J34 engines, the X-3 made its maiden flight on 20 October 1952. The programme was terminated in May 1956, after 20 flights had been carried out by NACA.

Fairey Delta 2 supersonic research aircraft

The first successful British convertiplane and first large VTOL transport was the Fairey Rotodyne, the prototype of which made its first flight on 6 November 1957. With accommodation for 40 passengers and a crew of two, the prototype Rotodyne Y was powered by two Napier Eland turboprop engines which were mounted below the fixed wings and drove tractor propellers. The large rotor mounted above the fuselage was driven by pressure-jets at the blade tips when required to be powered. In operation, the main rotor was usually powered for takeoff and landing, but was allowed to auto-rotate in normal horizontal flight, the forward propulsion of the aircraft and much of the lift then coming from the turboprop engines and wings respectively. The first transition from vertical take-off to horizontal flight was made on 10 April 1958, and although potential operators were enthused with the Rotodyne, the project was cancelled under government economic cutbacks in 1962.

The first British aircraft to be fitted with a rocket and a turbojet engine was the Saunders-Roe SR53 experimental interceptor, which flew for the first time on 16 May 1957. Although a developed production interceptor was expected to have a maximum speed above Mach 2·4, the type never entered service as its role was expected to be carried out in the future by surface-to-air missiles.

The fastest manned aircraft ever flown, and the aircraft flown to the greatest altitude, was the North American X-15A. The first powered flight by an X-15A was made on 17 September 1959, after air-launch from a B-52 carrier-plane. The aircraft itself was made mostly of titanium and stainless steel, the airframe being covered with Inconel X nickel alloy steel to withstand temperature of −300° to +1200°F (−184 to +650°C). After initial flights with an interim

Nord Griffon II with ramjet engine

rocket-engine, the X-15 was fitted with a Thiokol XLR99-RM-2 liquid-propellant rocket motor, giving a thrust of 57 000 lb (25 850 kg). In all, 199 flights were made by the three X-15As, the successive maximum speeds attained being listed on page 217. The greatest altitude reached was 354 200 ft (107 960 m), on 22 August 1963. The very successful X-15 programme came to an end in 1968.

The world's first fully-successful experimental V/STOL fighter was the Hawker Siddeley P1127. The first of two prototypes made its initial tethered hovering flight on 21 October 1960. In September 1961 the first transition flights from vertical to horizontal were made, and in 1963 the type was tested on board HMS *Ark Royal*. Several other development and evaluation models were built, with the name Kestrel, and were subsequently developed into the Harrier. Power for the Kestrel was provided by one Bristol Siddeley Pegasus 5 vectored-thrust turbofan engine of 15 200 lb (6895 kg) thrust.

The largest research aircraft ever built was the North American XB-70A Valkyrie, which was designed originally as a Mach 3 strategic bomber for the USAF but modified subsequently into an aerodynamic test vehicle. Two Valkyries were built, the first flying on 21 September 1964. Each had large delta wings, with

Fairey Rotodyne VTOL transport

hydraulically-drooping wingtips and 12 elevons and canard foreplanes. Power was provided by six General Electric YJ93 turbojet engines, each giving a thrust of 31 000 lb (14 050 kg) with afterburning. Mach 3 was achieved for the first time on 14 October 1965. Later, one of the Valkyries was destroyed after colliding with its accompanying chase fighter. The programme was completed in 1969 after more than 70 flights had been made by the surviving aircraft.

The Mikoyan Analogue was a MiG-21 fighter fitted with a scaled-down version of the original wings designed for the Tu-144 supersonic airliner, for aerodynamic testing.

The West German Dornier Do 31E was an experimental V/STOL transport aircraft, with

North American X-15 being air-launched from its B-52 mother-plane

Post War—Research continued on page 192

SPAD XIII 1719, *flown by Lieutenant Gervais Raoul Lufbery of Escadrille SPA124 – the Escadrille Lafayette – during the winter of 1917–18. With seventeen confirmed victories Lufbery was third in the roll of American aces. Born in France of French parents who later emigrated to America, Lufbery travelled widely as a young man; when war broke out he joined the French Foreign Legion and obtained a quick transfer to the Service Aéronautique. Having successfully applied for pilot training, he received his Brevet on 29 July 1915, and was posted to Escadrille de Bombardement VB 106. In May 1916 he was transferred to the Escadrille Lafayette, and had scored five victories by early October. His subsequent career brought him promotion and many decorations, including the first award of the British Military Cross to an American. With America's entry into the war Lufbery was transferred, with the rank of Major, to the US Air Service, and was subsequently given command of the famous 94th Aero Squadron. He was killed on 19 May 1918; flying a Nieuport 28, he was attacking an Albatros two-seater reconnaissance aircraft over the 94th Squadron's base at Toul when his machine burst into flames. Lufbery fell from the burning aircraft at an altitude of 6000 ft – whether deliberately or accidentally will never be known. The Sioux Indian head insignia was the badge of SPA124; the swastika was Lufbery's good luck symbol.*

SPAD XIII *flown by Maggiore Francesco Baracca as commanding officer of the 91a Squadriglia of the Italian Aeronautica del Regio Esercito during the winter of 1917–18. Italy's leading ace with thirty-four confirmed victories, Baracca had qualified as a pilot in 1912, and had flying experience with many types of Italian, French and Belgian aircraft by the time Italy declared war on Austria in May 1915. His first victory was gained on 7 April 1916; leading a dawn patrol of Nieuport 11 scouts from the 70a Squadriglia, he shot down an Aviatik two-seater near Medea. His fifth victory came on 25 November 1916, and from this date onwards he had his famous prancing horse insignia painted on all his aircraft. He continued to score steadily, including two 'doubles' – on 21 and 26 October 1917 – and was heavily decorated. He failed to return from a ground-strafing mission near Montello on 19 June 1918, and his body and burned-out aircraft were not located until after the Austrian retreat; various theories about his death gained currency, but it seems most probable that he was killed by ground fire from some infantryman crouching in a trench.*

Hanriot HD-1 *flown by Lieutenant Jan Olieslagers of the 1ere Escadrille de Chasse, Aviation Belge Militaire. A pre-war motor-cycle racing star and a pilot since 1909, Olieslagers became one of the great names of Belgian aviation. He served with distinction in Belgium's first all-fighter squadron alongside such aces as Willy Coppens and André de Meulemeester; although his official score was six victories, his contemporaries in the 1ere Escadrille maintained that he scored many more kills behind the German lines, where ground confirmation was seldom possible. Olieslagers died in Antwerp in 1942; the thistle insignia on his aircraft was the badge of the 1ere Escadrille.*

Morane-Saulnier MS 5 *scout flown during 1915 by Staff Captain Alexander Alexandrovitch Kazakov as commander of the XIX Corps Air Squadron, Russian Imperial Air Service. Kazakov is acknowledged as Russia's leading First World War ace with seventeen confirmed victories, but unofficial estimates put his true score at thirty-two. He conceived the idea of fixing a steel grapnel on a length of cable to his unarmed Morane scout, and flying low over enemy aircraft to tear away wings, control surfaces or flying wires. He scored his first victory on 18 March 1915, when he brought down an Albatros two-seater near Gusov partly by means of his grapnel, partly by ramming it with his undercarriage. A quiet and deeply religious man, Kazakov went on to command the 1st Fighter Group comprising four squadrons, and to receive sixteen decorations including the British DSO, MC, and DFC. During the British intervention in the Russian Civil War he was attached to the RAF, flying Sopwith Camels. Shortly after the announcement that British forces were to be withdrawn from the campaign, Kazakov died in a rather ambiguous and inexplicable flying accident on 3 August 1919.*

Phönix D I *flown by Oberleutnant Frank Linke-Crawford as commander of Fliegerkompagnie 60J, Austro-Hungarian Imperial Luftfahrtruppen; this squadron was based at Feltre on the Piave front during the winter of 1917–18. With an official score of twenty-seven to thirty, Linke-Crawford stands third in the roll of Austro-Hungarian aces; little is known of his personal life, beyond the fact that he was a rather earnest and reserved young man who earned promotion at a rate considered phenomenal in the ultra-conservative Imperial forces. He was shot down by Captain J Cottle of 45 Squadron, RAF on 31 July 1918 near Montello.*

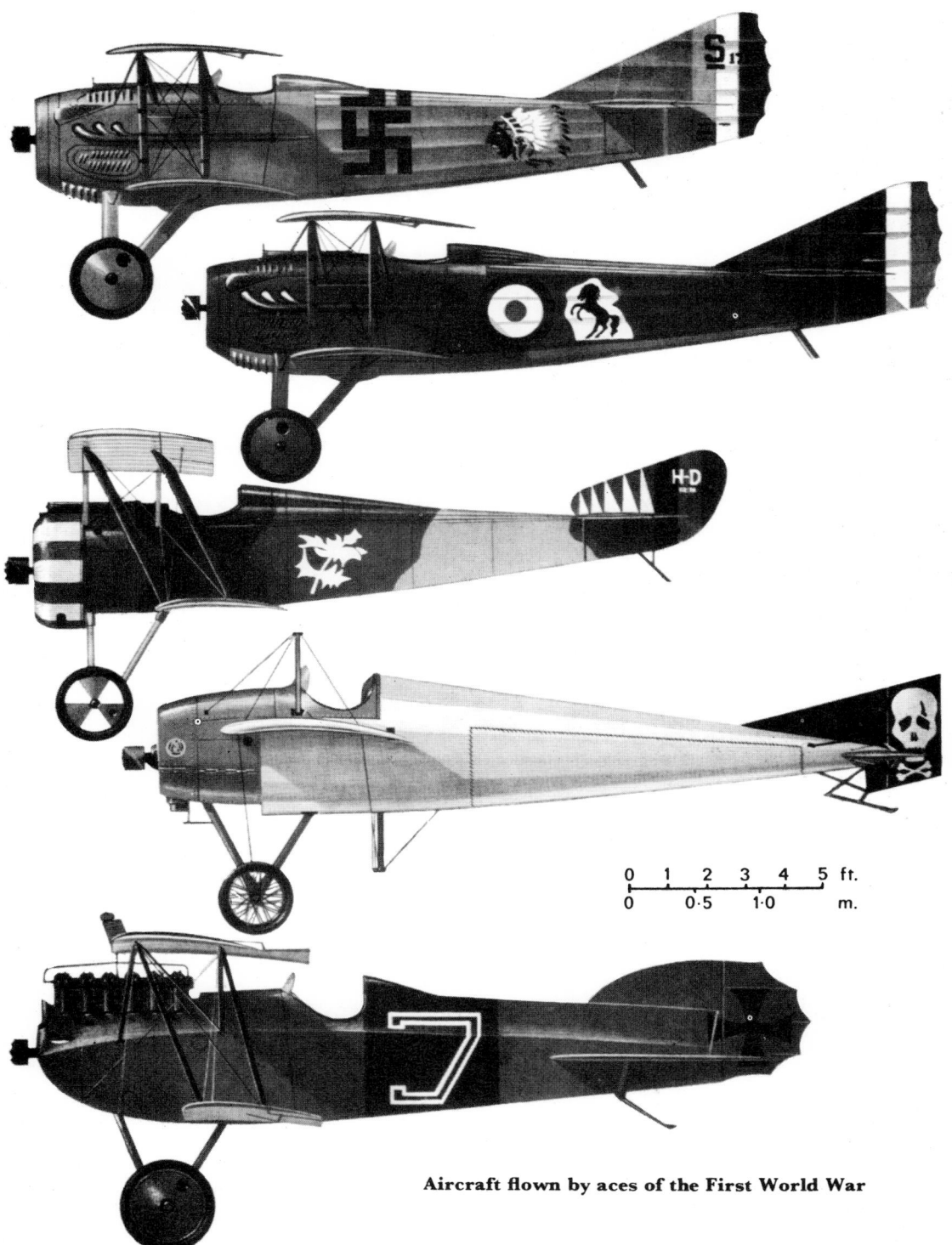

Aircraft flown by aces of the First World War

accommodation for freight or 34 troops. It was powered in forward flight by two Rolls-Royce Pegasus vectored-thrust engines, mounted under the wings. Vertical lift was achieved via eight Rolls-Royce RB162-4D turbojet engines, housed in two wingtip pods. Two flight-test aircraft were constructed, the first flying on 10 February 1967. Transition from vertical to horizontal flight was achieved by the second aircraft on 16 December 1967, and many more flights were made before the programme was brought to a close in 1970.

The last known flight of a rocket-powered aircraft was made by the Martin Marietta X-24B lifting-body research vehicle on 23 September 1975. The X-24B was basically the earlier X-24A rebuilt to have a longer pointed fuselage with the flat side of its triangular cross-section on the bottom instead of on the upper surface as with the X-24A. Its first unpowered flight was made on 1 August 1973, followed by five further unpowered and 14 powered flights, the latter using the installed Thiokol XLR11 turbo-rocket engine. Two small rocket motors could be fitted as a landing aid. The purpose of the X-24 research programme was to provide data on the flight characteristics of vehicles flown in orbit and then brought back through the atmosphere to land as conventional aircraft, a system used on the Space Shuttle Orbiter vehicle.

Most advanced helicopter built solely for testing rotor systems is the Sikorsky S-72 or Rotor System Research Aircraft (RSRA). The contract for two such helicopters was placed with Sikorsky in 1974 by NASA, who ordered also a pair of General Electric TF34 turbofan engines, in pods, for optional mounting on the sides of the fuselage of one of the helicopters, for high cruising speed research, and two pairs of 41 ft 10 in (12·75 m) span wings. Each powered by two General Electric T58 engines of 1400 shp each, the S-72s were delivered for use by NASA and the US Army in 1977, as test vehicles for new rotor systems under development or projected for future use. One of the most important features of the design is the ability of the S-72 to fly satisfactorily as a conventional aeroplane, relying for lift and thrust on its wings and turbofan engines only. This enables experiments to be carried out on rotors that would be too small to keep the helicopter airborne. It also provides a safety back-up system if a new rotor under test should malfunction. Maximum speed of the S-72 is about 345 mph (555 km/h).

Early Cierva Autogiro, based on an Avro 504 airframe

SECTION 11

ROTORCRAFT

The first documented reference to the possibility of sustaining or propelling upwards a vehicle by means of rotating surfaces is attributed to Leonardo da Vinci (1452–1519), whose design sketches for such are believed to have originated in about the year 1500. Leonardo was otherwise devoted to the concept of flapping wings (i.e. the ornithopter) to achieve forward flight, and he was not aware of the lifting characteristics of aerofoils, nor was he acquainted with the properties of the propeller. As a result his design for a helicopter was based strictly on an 'air screw' – literally a rotating helical wing which would 'screw' its path upwards through the air.

Numerous attempts to evolve models of helicopters followed during the next four centuries, culminating in the unmanned models of W H Phillips who, in 1842, succeeded in launching a steam-driven craft whose rotating *blades* were propelled by tip jets.

It is perhaps useful here to interpose simple definitions of the helicopter and autogyro. Basically a helicopter achieves vertical flight by means of aerodynamic lift from rotor blades which are rotated under power; to eliminate torque (i.e. to prevent the fuselage of the aircraft from spinning uncontrollably on the axis of the rotor), either coaxial rotors, balanced sets of rotors or small tail-mounted rotors are geared to the power plant. Forward flight is achieved by tilting the rotor 'disc' so that its resulting thrust provides a degree of propulsion as well as lift.

An autogyro, on the other hand, is rather nearer to a conventional aeroplane in that forward motion is achieved by a conventional engine (either jet or piston engine-driven propeller); as forward motion is achieved the freely rotating rotor blades provide lift as aerofoils, enabling the autogyro to perform short, steep take-offs and landings.

The first helicopters to fly were small models powered by the string-pull method, the first of which appeared in the 14th century. The first picture of one of these models appears in a painting of 1460.

The first self-propelled model helicopter appears to have been that demonstrated on 28 April 1784, in France, by Launoy and Bien-venu. It consisted of a stick with a two-blade propeller at each end. The model was powered by a bowdrill arrangement; as the string of the bowdrill unwound the propellers contrarotated. It was on this model that Sir George Cayley based his model helicopter in 1796, using a similar bowdrill arrangement but powering two four-blade rotors made from feathers.

Igor Sikorsky

Sir George Cayley's 'Aerial Carriage' designed in 1843 was perhaps the first attempt at a convertiplane. It had four circular wings, in pairs, mounted on outriggers from the boat-like wheeled fuselage. When the rotating wings were needed to provide lift, they were designed to open out into eight-bladed propellers. Forward propulsion was by two rear-mounted propellers.

The first helicopter to lift a man from the ground was built by the Breguet brothers of France in 1907. Although the craft lifted off the ground at Douai, France, on 29 September that year, it did not constitute a free flight as four men on the ground steadied the machine with long poles which, while not contributing to the aircraft's lift, constituted a form of control restriction. Power was provided by a 50 hp Antoinette engine.

The first true free flight by a man-carrying helicopter was performed by Paul Cornu in his 24 hp Antoinette-powered twin-rotor aircraft near Lisieux, France, on 13 November 1907.

Neither of the above helicopters incorporated cyclic pitch control (hence the problem overcome by Breguet in using external control stabilisation) although G A Crocco had in 1906 suggested its necessity. It was J C H Ellehammer, the Danish pioneer aviator, who first produced in 1912 a helicopter (of only limited application) which incorporated cyclic pitch control.

The first helicopter to be built by Igor Sikorsky (who later became one of the greatest helicopter designers in the world) was a twin contra-rotating rotor machine which appeared in 1910. Each rotor consisted of two long-chord wing-type planes and an aerodynamically-shaped spar; the fuselage was a simple upright structure designed to hold an engine to drive the rotor via a belt. The whole apparatus was mounted on rails. However, the helicopter did not fly and it was to be a quarter of a century before Sikorsky would fly successfully in a helicopter of his design.

The first helicopters to successfully demonstrate cyclic pitch control were those designed and constructed by the Marquis de Pescara, an Argentinian, in France and Spain between 1919 and 1925. Although demonstrating this feature successfully, these aircraft were directionally unstable as the result of inadequate torque counteraction.

The first successful flight by a gyroplane, the C4 (commercially named an *Autogiro*) was accomplished by the Spaniard Juan de la Cierva at Getafe, Spain, on 9 January 1923.

The first two-seat Autogiro in the world was the Cierva C6D which was first flown by F. T. Courtney at Hamble, England, on 29 July 1927. Don Juan de la Cierva, the Spanish inventor, became **the first passenger in the world to ride in a rotating-wing aircraft** when he was taken aloft the following day in this Autogiro.

The first rotating-wing aircraft to fly the English Channel was the Cierva C8L Mark II (*G-EBYY*) Autogiro flown by Don Juan de la Cierva with a passenger from Croydon to Le Bourget on 18 September 1928.

The first helicopter to fly successfully, although it was never developed beyond the prototype stage, was that built by Louis Breguet and flown in 1936. It featured contra-rotating rotors, an open framework fuselage, an aeroplane-type tail unit and a wide-track main undercarriage with additional nose and tailwheels. This helicopter flew for over an hour during which it covered 27 miles (43·5 km).

The first entirely successful helicopter in the world was the Focke-Wulf Fw 61 twin-rotor helicopter designed by Professor Heinrich Karl Johann Focke during 1932–4. The first prototype Fw 61V1 (*D-EBVU*) made its first

Focke-Wulf Fw 61

Igor Sikorsky flying his VS-300, from which all modern 'single-rotor' helicopters have derived

free flight on 26 June 1936 and was powered by a 160 hp Siemens-Halske Sh 14A engine. This aircraft, flown by Ewald Rohlfs in June 1937, established a world's closed-circuit distance record for helicopters of 76·025 miles (122·35 km) and a helicopter endurance record of 1 h 20 min 49 s. On other occasions it set up an altitude record of 11 243 ft (3427 m) and a speed record of 76 mph (122 km/h). It gave a flying demonstration in the Berlin Deutschland-Halle during 1938 in the hands of the famous German woman test pilot Hanna Reitsch.

The first helicopter to go into limited production was the Focke-Achgelis Fa 223. The experimental Focke-Wulf Fw 61 (see above) was not commercially exploited, being too heavy structurally to carry a payload. Instead, a commercially developed derivative, the Fa 266 Hornisse appeared in 1939 as a prototype six-seat civil transport helicopter. This was **the first real transport helicopter.** The Fa 266 first made a free flight in August 1940 and was redesignated Fa 223 Drache, by which time it had changed into a military helicopter. By 1942 the Fa 223 was ready for operational trials although only two examples had flown because of Allied bombing. Because of the bombing, the factory had moved from Bremen to Laupheim and eventually finished up in Berlin. By the end of the war only a small number of helicopters had flown, three of which were used for transport duties by Luft-transportstaffel 40.

The first successful helicopters to be designed outside Germany were those of the Russian-born American Igor Sikorsky. His first successful helicopter was the VS-300. Powered by a 75 hp engine, it featured full cyclic pitch control and achieved tethered flight on 14 September 1939 in America. Its first recognised free flight was made on 13 May 1940. In May 1941

a 90 hp engine was fitted and the VS-300 set up a new endurance record of 1 h 32 min 26 s. By 1942, after further improvement, the VS-300 became established as the first successful and practical single-rotor helicopter.

The Focke Achgelis Fa 330 Bachstelze was a single-seat gyro-kite designed in Germany in early 1942 and built by the Weser-Flugzeugbau at Hoykenkamp, near Bremen. It was designed to be carried on board the ocean-going IX type submarine and, when needed for observation duties, to be towed behind the surfaced submarine at about 400 ft (122 m) altitude, giving the pilot about 25 miles (40 km) vision. However, U-boat crews disliked the Bachstelze and so it was not often used.

The first helicopter designed and built for military service was the Sikorsky XR-4 which first flew on 13 January 1942 and was delivered to Wright Field, Dayton, Ohio, that year for military evaluation. As a result of these trials, machines of a small development batch were used for limited service and training in 1944 and 1945, R-4s being the first helicopters to fly in Burma and Alaska, first to be tried on board ship (1943), and the first to be flown by the British Fleet Air Arm.

The first jet-driven helicopters in the world were the Doblhoff/WNF 342 (V1–V4) helicopters, built in the suburbs of Vienna between 1942 and 1945 by the Wiener Neustadter Flugzeugwerke (WNF). The principle on which jet power was produced was that compressed air, provided from a compressor driven by a piston engine, was mixed with fuel and channeled through the three hollow rotor blades and burnt at the rotor tips in combustion chambers. The first helicopter, the V1, first flew in the spring of 1943 but was slightly damaged in 1944 by bombing raids. The V2 helicopter (the V1,

First helicopter production line – R-4s being manufactured at Bridgeport, Connecticut, USA

modified) was then completed, followed by the V3, which destroyed itself by vibration. The final version, the V4, flew well, but work had to be stopped in 1945.

The first helicopter to go into full production and service was the Sikorsky R-4B, at Bridgeport, Connecticut, in 1944. However, in 1943 the American Navy had taken a small number of early R-4s to sea to find out how useful the helicopter could be in detecting submarines, with the view of using them on merchant ships, etc. Unfortunately, although successful flights were made at sea, the R-4 was not sufficiently developed for this work.

The first successful tandem twin-rotor helicopter to be put into production was the Piasecki PV-3 (US Navy designation HRP-1) which first flew in March 1945. It was powered by a 600 hp Pratt & Whitney R-1340 Wasp

engine and could fly at 120 mph (193 km/h). Designed to carry ten persons, six stretchers or cargo, the first production HRP-1 was completed on 15 August 1947. Versions of this helicopter served with the US Navy, Marine Corps, Coast Guard and Air Force.

The first helicopter to cross the English Channel was a Focke-Achgelis Fa 223 (No 14) which arrived at Brockenhurst, Hampshire, in September 1945, piloted by a German crew of three. Fa 223 No 14 was first flown in July 1943 and with No 51 was confiscated by the Ameri-

Sikorsky R-4B, taking off from US 10th Air Force airfield in Burma, January 1945

Post-war tests of a captured Focke-Achgelis Fa 330 rotor-kite. This unpowered aircraft was intended to be towed behind U-boats to locate shipping targets 'over the horizon' (Royal Aircraft Establishment)

Bell Model 47-B in Helicopter Air Transport livery

cans in May 1945. No 14 was destroyed in October 1945 during evaluation trials.

The first-ever Type-Approval Certificate awarded for a commercial helicopter was for the Bell Model 47 on 8 March 1946; this aircraft made its first flight on 8 December 1945 and provided the design basis for a family of Bell helicopters which has continued in production for more than thirty years. The Model 47 is now built only in Italy, by Agusta.

The world's first scheduled helicopter service was inaugurated on 1 October 1947 by Los Angeles Airways (LAA), using a Sikorsky S-51. LAA had been given a three-year temporary certificate for mail carrying, on 22 May 1947, by the CAB.

The first helicopter built in Great Britain to enter service with the RAF was the Sikorsky-designed Westland-Sikorsky Dragonfly (the S-51 built under licence by Westland Aircraft Ltd, Yeovil, England). The first West-

land-built S-51 was for commercial use and flew in 1948. The RAF's first helicopter, a Dragonfly HC Mark 2 (*WF308*) powered by an Alvis Leonides engine, was delivered in 1950, and subsequent aircraft equipped No 194 (Casualty Evacuation) Squadron, **the RAF's first helicopter squadron,** on 1 February 1953.

The first helicopter mail service in Great Britain was inaugurated on 1 June 1948 by British European Airways, flying a Westland-Sikorsky S-51. Based at Peterborough it served Norwich, Great Yarmouth and Kings Lynn.

The first experimental night helicopter service was inaugurated on 20 February 1949 by British European Airways Corporation, flying Westland-Sikorsky S-51 *G-AKCU*. The service, from Westwood, Peterborough, to Norwich, became regular from 17 October 1949.

The first British-designed and -built production helicopter was the Bristol Sycamore

Bristol Sycamore, first British production helicopter

which first flew on 27 July 1947 and entered service with both the Army and Air Force. The Army versions were the HC10 ambulance and HC11 communication helicopters, the latter initially flying on 13 August 1950 and being delivered from 29 May 1951. The first RAF version, HR12, was sent to St Mawgan for trials on 19 February 1952.

The first helicopter to have the engine mounted in the nose of the fuselage was the Sikorsky S-55, which first flew on 7 November 1949. This arrangement left the cabin area free for passengers or cargo.

The first scheduled passenger helicopter service was inaugurated on 1 June 1950 by British European Airways, flying Westland-Sikorsky S-51s. The service was between Cardiff and Speke airport, Liverpool. The service ended on 31 March 1951 because of low demand, after carrying 819 passengers.

The first war in which helicopters were widely used was the Korean War of the early 1950s. Used for such roles as rescue, ambulance, assault, logistic support and flying-crane transport, helicopters opened a completely new era in warfare. During the three years of war, over 23 000 casualties were carried by helicopter to field hospitals; making possible the smallest percentage of wounded to die in recorded military history.

The first international helicopter service was inaugurated on 1 September 1953 by SABENA, flying Sikorsky S-55s. Services included flights from Brussels to Rotterdam, Lille and Maastricht.

The autogyro that has been produced in greater numbers than any other non-military rotorcraft is the Bensen B-8M Gyrocopter, the prototype of which first flew on 6 December 1955. More than 7000 examples of the B-8M had been built by late 1974 and the current version is usually powered by a 72 hp McCulloch Model 4318E engine which gives it a maximum speed of 85 mph (137 km/h).

The first rotating-wing aircraft to be literally a 'flying-boat' was the Bensen Model B-8B Gyro-Boat. Designed as a variant of the Gyro-Glider, this aircraft was basically a small dinghy that was fitted with a two-blade free-turning rotor and outer stabilising floats (in later models). The Gyro-Boat would take off after being towed by a motorboat to a speed of 23 mph (37 km/h). The prototype first flew on 25 April 1956 and many such aircraft were built.

The Kamov Ka-22 Vintokryl was a large twin-engined convertiplane. It was first shown publicly in 1961 and was designed to carry over 80 passengers. Powered by two Soloviev D-25V engines, it flew to a height of 8491 ft (2588 m) on 24 November 1961 at Bykovo, USSR, so gaining world records for altitude, payload-to-height (2000 m), and greatest load carried in the E2 class (convertiplanes). On this flight a 36 283 lb (16 458 kg) load was carried to the above altitude, so setting up records for 1000 kg, 2000 kg, 5,000 kg, 10 000 kg and 15 000 kg loads. This convertiplane also set a world record for speed in a straight line when, on 7 October 1961, it flew at 221·4 mph (356·3 km/h) over a 15/25 km course.

Perhaps the most unusual helicopters ever produced, and certainly the lightest, are the Seremet strap-on helicopters, the first of which (the WS1) was tested in 1961. The latest is the WS8, flight testing of which began in 1975. It consists basically of a small engine mounted on a back-pack, the latter also supporting the twist-grip extended arm controls, the two-blade rotor and the small tail-rotor which is situated on the end of a rearward extending single thin boom.

The first specially designed combat helicopter to go into large-scale service was the Bell Model 209 HueyCobra. First flown on 7 September 1965, six months after its development was started, the HueyCobra is a two-seat armed combat helicopter, designed for ground attack and helicopter escort duties. It can fly at over 200 mph (322 km/h) and can carry guns, grenade launcher, rockets and missiles. One of the main features of the HueyCobra is that its fuselage is only 38 in (0·965 m) wide, making it a difficult target to hit by ground fire and easy to conceal with small camouflage nets or under trees.

The greatest distance covered in a straight line by a helicopter at the time of writing is 2213 miles (3561·6 km), performed by a Hughes OH-6A light observation helicopter between 6 and 7 April 1966. The pilot was R G Ferry who flew from California to Florida, USA, non-stop.

The fastest autogyro over a 3 km straight course and over 100 km and 500 km closed circuits at the time of writing is the Wallis WA-116. Originally powered by a 72 hp modified McCulloch 4318 engine, a WA-116/Mc gained a class E3 (Autogyro) height record of 15 220 ft (4639 m) on 11 May 1968, while piloted by Wing Commander K H Wallis, and on 12 May 1969 gained the 3 km speed record with a speed of 111·225 mph (179 km/h). One Wallis WA-116, re-engined with a Franklin 2A-120-2B and modified further by the fitting of an eleven Imperial gallon (50 litre) internal fuel tank and an 8 Imperial gallon (36 litre) external tank, gained new E3 and E3a world records by flying non-stop for 416 miles (670 km) in a 100 km closed circuit in July 1974. This flight also set the 100 km speed record at 81·19 mph (130·7 km/h) and the 500 km speed record at 78·38 mph (126·14 km/h).

The highest take-off and landing ever recorded by a helicopter (at the time of writing) was that made by an Aérospatiale Lama in 1969 during demonstration flights in the Himalayas. With two persons on board, the helicopter landed at 24 600 ft (7500 m).

The largest helicopter in the world in 1977 was the Russian Mil Mi-12 twin-rotor aircraft with an overall span (over the rotor tips) of 219 ft 10 in (67 m), and fuselage length of 121 ft 4½ in (37 m). It is powered by four 6500 shp Soloviev D-25VF turboshaft engines. First indication of the existence of this enormous helicopter to reach the West was in 1969 when Russia submitted for ratification the lifting of a record payload of 68 410 lb (31 030 kg) to an altitude of 9682 ft (2951 m) – approximately the loaded weight of an Avro Lancaster bomber of the Second World War. On 6 August 1969 the Mi-12 beat its own record by lifting a payload of 88 636 lb (40 204·5 kg) to a height of 7398 ft (2255 m).

The absolute height record for a helicopter is held by an Aérospatiale Lama with an altitude of 40 820 ft (12 442 m), achieved on 21 June 1972. The pilot was Jean Boulet.

The newest assault helicopter to be produced in the USSR is the Mil Mi-24 (**NATO code name Hind**). It is approximately 55 ft 9 in (17 m) long and has auxiliary wings, each with three weapon attachment points for rocket pods and missiles. A flexible 12·7 mm machine gun is mounted in the nose. Mi-24s are operational in the USSR and in East Germany.

One of the few helicopters to be designed for home-construction, the RotorWay Scorpion Too is a two-seat lightweight sporting helicopter, powered by a 140 hp Vulcan V-4 engine. Basically of steel-tube construction, with a removable fibreglass cabin, it has a cruising speed of 75 mph (121 km/h).

The military force with the greatest number of helicopters on strength is undoubtedly the US Army, which has nearly 9000 helicopters of all kinds.

Sikorsky S-67 Blackhawk, world's fastest helicopter, which achieved 220·888 mph (355·485 km/h)

Mil Mi-24, an assault helicopter which can carry a squad of eight troops and keep down the heads of enemies in the drop-zone with a machine-gun and underwing rockets and missiles

World's smallest aeroplane, the Stits Skybaby.
Powered by an 85 hp Continental piston-engine, it
could fly at 185 mph (298 km/h), but spanned only
7 ft 2 in (2·18 m) (Howard Levy)

SECTION 12

FLYING FOR SPORT AND COMPETITION

In realms of commerce and in times of war it is perhaps often overlooked that flying originally approached maturity before the First World War largely through the enthusiasm and courage of a small band of philanthropic sportsmen who sought satisfaction through a precarious recreation, and the measure of whose success lay in their ability to fly further, faster or higher than their colleagues at home and abroad. The very nature of their sport, eliminating as it did the frontiers that had divided nation from nation, brought airmen of different countries closer together, so that almost at once international competition was keen – yet characterised by something of a universal cameraderie. So powerful was this sense of *entente* between all aerial sportsmen that when, inevitably, the Kaiser's War split nations asunder and they were placed under orders to do battle with one another, there was a reluctance at first to so prostitute their new-found sport by turning their frail craft into machines of destruction. One has only to attend an international flying meeting 60 years later to realise that the old *entente* has not altogether disappeared. . . .

This section spans the scope of flying by pilots who, perhaps supported only by limited money, indulged their whims in the private ownership of an aeroplane, and also by pilots who, by their prowess (or perhaps in spite of their lack of it) are encouraged to compete against each other either in organised or in informal sporting events. While of course there have been the classic annual spectacles – the Schneider Trophy, the King's Cup Air Race and the American National Air Races – there have also been great once-only competitions, sponsored by big finance, like the Australia Race of 1934 and the Transatlantic Race of 1969. All have contributed greatly both to the progress of aviation and to the better understanding of what, after all, aviation has to offer: the drawing together of nation to nation by time, distance and competition.

The first major prize to be offered in Great Britain for a feat performed in an aeroplane was the £1000 offered by the *Daily Mail* for the first aeroplane flight across the English Channel.

This was won by the Frenchman, Louis Blériot, who crossed from France to England on Sunday, 25 July 1909 (see page 42).

The World's first international air meeting was held at Reims, France, and opened on 22 August 1909. Promoted and financed by the French Champagne industry, which also offered a number of generous prizes, no fewer than thirty-eight entries were received. Among the aeroplanes which assembled were seven Voisins, six Wright biplanes, five Blériot monoplanes, four Henry Farman biplanes, four REPs, three Antoinettes, two Curtisses, one Santos-Dumont, one Breguet and one Sanchis. Competition during the meeting involved a number of speed, duration and distance events, and Henry Farman established world records for duration and distance by flying 111·8 miles (180 km) in a closed circuit in 3 h 4 min 56·4 s. His prize money totalled 63 000 francs. Fastest speed was recorded by Louis Blériot who, flying his Blériot XII monoplane, won the 10 km speed contest at 47·8 mph (76·99 km/h). *See also 'Pioneers'*.

The first major British aviation meeting was that held at Doncaster, from 15–23 October 1909. A total of nine aircraft took part, comprising five Blériot XI monoplanes, a Farman and a Voisin biplane, Captain W G Windham's tractor monoplane and the colourful S F Cody's British Army Aeroplane No 1.

The first prize of £10 000 to be offered in Great Britain for an aeroplane flight was again offered by the *Daily Mail* for the first pilot to fly an aeroplane from a point within 5 miles of the newspaper's London offices to a point within 5 miles of its Manchester offices. This prize encouraged Louis Paulhan and Claude Grahame-White to compete in their Henry Farman biplanes, success finally attending the efforts of the former on Thursday, 28 April 1910. The event, which ultimately developed into a race (with Grahame-White resorting to an epic night flight, the first in Europe), fired the imagination of the whole country as supporters and officials hired special trains to follow the progress of the aviators, and reports were flashed to HM King Edward VII who was abroad at the time.

While Paulhan (above) waited for the dawn at Lichfield, Grahame-White made his epic night flight

The first prize won by an Englishman flying in America went to Claude Grahame-White who, between 6 and 12 September 1910, won a £2000 prize offered by the *Boston Globe* by flying 33 miles (53 km) cross-country in 40 min 1·6 s. At the same Boston meeting Grahame-White also gained four other first prizes and three seconds, so that his total prize-money amounted to £6420 for the week.

The first Gordon Bennett international aeroplane race was flown at the end of October 1910 at Belmont Park, New York, and attracted teams from the USA (Messrs Brookins, Drexel and Moisant), France (Messrs Latham and Leblanc) and Great Britain (Messrs Grahame-White, Ogilvie and Radley). On 29 October the race was won by Claude Grahame-White, flying a Blériot monoplane powered by a 100 hp Gnome rotary engine, who completed the 62·2 mile (100 km) course in 1 h 1 min 4·74 s. Second place was taken by Moisant, also flying a Blériot.

The longest straight-line distance flight from the UK into Europe in 1910, winning the Baron de Forest prize of £4000, was achieved by T O M Sopwith on 18 December. Flying a Howard Wright biplane, Sopwith flew 169 miles (272 km) from Eastchurch to Thirimont, Belgium, in 3 h 30 min.

The emphasis placed upon financial reward for competition in the air during the early years of aviation may, by modern definition, tend to detract from the amateur status of the pilots of that time, yet it was the nature of their sport and the absence of national support which threatened to retard their progress by its very cost. It was the far-sighted sponsorship by the Press and by such persons as Archdeacon and the Baron de Forest which not only ensured the survival of competitive aviation, but also spurred European aviation to the efforts which very quickly overtook those of the pioneering Americans. For example, Claude Grahame-White won in prize-money during 1910 a total of £10280 – much of which was applied to the support of the important Grahame-White Flying Schools.

The first truly international air race (point-to-point) held in Europe was the Circuit of Europe which started on 18 June 1911, over the route Paris – Reims – Liège – Spa – Liège – Verloo – Utrecht – Breda – Brussels – Roubaix – Calais – Dover – Shoreham – London – Shoreham – Dover – Calais – Amiens – Paris. The field included eight Moranes, seven Deperdussins, six Blériots, three Sommers, three Caudrons, three Henry Farmans, two Maurice Farmans, two Bristols, two Voisins, two Astras, and one each of Antoinette, Barillon, Bonnet-Lab, Danton, Nieuport, Pischoff, REP, Tellier, Train, van Meel and Vinet. The race was won after nineteen days on 7 July by Lieutenant de Vaisseau Conneau flying a Blériot, followed by Roland Garros also in a Blériot. Prize-money totalled £18300, and only nine aeroplanes completed the course.

The first 'Round-Britain' air race was sponsored by the *Daily Mail*, which newspaper presented a £10000 prize for a race which started from Brooklands on 22 July 1911. The course, which was flown in five days, followed the route Brooklands – Hendon – Harrogate – Newcastle – Edinburgh – Stirling – Glasgow – Carlisle – Manchester – Bristol – Exeter – Salisbury Plain – Brighton – Brooklands, a distance of 1010 miles (1625 km). Only two French aircraft completed the course in the specified time and the race was won by the French naval officer Lieutenant de Vaisseau Conneau in a Blériot XI, in a time of 22 h 28 min. Jules Vedrines came second, and Samuel Cody completed the course three days late.

The first Schneider Trophy Contest (more correctly titled 'La Coupe d'Aviation Maritime Jacques Schneider'), was included as one item of the second international Hydro-aeroplane Meeting held at Monaco during the two weeks beginning 3 April 1913. It created initially little interest, with only seven entries for the first contest, reduced to four starters after the eliminating trials. The course consisted of twenty-eight 10 km laps and this 1913 contest, flown on 16 April, was won by Maurice Prévost flying a 160 hp Gnome-powered Deperdussin. This pilot was under the impression that he should alight and taxi across the finishing line, but in fact this was invalid and it was required that he should fly the aircraft across the line. He was accordingly sent off again to complete a further lap which, including a period spent ashore deciding whether to continue or not, added about an hour to his time, so that his average speed is recorded in the official results as being 45·75 mph (73·63 km/h). Second was late-starter Roland Garros,

The Schneider Trophy

flying a Morane-Saulnier monoplane powered by an 80 hp Gnome engine.

The first non-stop flight of a powered aircraft from England to Germany was achieved, on 17 April 1913, by Gustav Hamel flying a Military-type Blériot XI. The flight from Dover to Cologne took a time of 4 h 18 min.

The first major British competition for seaplanes was the *Daily Mail* Hydro-Aeroplane Trial, started on 16 August 1913. The regulations stated a specified course round Britain, involving a distance to be flown of 1540 miles (2478 km) by an all-British aircraft before 30 August. Four aircraft were entered, but Samuel Cody was killed in a crash at Laffan's Plain on 7 August. F K McClean withdrew his Short S68 due to engine trouble, and the Radley-England Waterplane was scratched for the same reason. This left Harry Hawker, accompanied by his mechanic H A Kauper, as the only contender. He

left the water at Southampton at 11.47 h, in a Sopwith three-seater tractor biplane which was powered by a 100 hp Green six-cylinder in-line engine. The route was from Southampton via Ramsgate, Yarmouth, Scarborough, Aberdeen, Cromarty, Oban, Dublin, Falmouth and back to Southampton. After an abortive attempt, which ended at Yarmouth owing to a cracked engine cylinder, Hawker took off again from Southampton on 25 August. He managed to fly round the course as far as Dublin when, just before alighting on the water, his foot slipped off the rudder-bar and the aircraft struck the water and broke up. The *Daily Mail* prize of £5000 was not awarded, but Hawker received £1000 as consolation.

The first British aeroplane to beat all comers in a major international competitive event was the Sopwith Tabloid. Designed as a small, fast biplane scout aircraft, it first flew in the autumn of 1913. Official tests at Farnborough on 29 November 1913 showed it had exceptional performance, with a maximum rate of climb of 1200 ft (366 m)/min and a maximum speed of 92 mph (148 km/h). Its outstanding competitive success was its victory in the second contest for the Schneider Trophy held at Monaco on 20 April 1914 when, equipped as a float-plane, the aircraft was flown by Howard Pixton over the 280 km course at an average speed of 86·78 mph (139·66 km/h). After completing the race, Pixton continued for two extra laps to establish a new world speed record for seaplanes at 86·6 mph (139·37 km/h) over a measured 300 km course.

The first major British aviation competition to be won by an American pilot was the third Aerial Derby, flown from Hendon in bad weather on 6 June 1914 and won by W L Brock in a Morane-Saulnier monoplane powered by an 80 hp Le Rhône engine. Completing the 94·5 mile (152 km) course at an average speed of 71·9 mph (115·7 km/h), Brock won the *Daily Mail* Gold Cup and Shell Trophy, as well as £300 in prize-money.

The first post-war aviation event to be held in Britain took place at Hendon on 7 June 1919, when Lieutenant G R Hicks won a cross-country handicap race flying an Avro 504K.

The first Schneider Trophy Contest to take place off the British coast was that flown at

Bournemouth on 10 September 1919. Fog turned the event into chaos, and it was eventually abandoned. Guido Janello of Italy completed eleven laps, but as there was doubt concerning one of his turning points it was not allowed to count as a victory. As a gesture, Italy was asked to organise the next event. This contest was the first in which an aircraft was involved that had been specially prepared by R J Mitchell, a 24-year-old recruit of the Supermarine Company.

In order to regain some of their country's lost leadership in aircraft design, businessmen in America in the early 1920s, with the co-operation of the military, put up cash prizes and trophies to promote high-speed air racing and advanced design. One of the first competitions was that organised by Ralph Pulitzer in 1920. Success was spectacular. By the time of the 1923 competition one of the entries, the Verville-Sperry R-3, appeared as a cantilever low-wing monoplane with a retractable undercarriage propeller spinner and neatly-cowled engine. In the same year the American Navy Curtiss CR-3 racer won the Schneider Trophy Contest, followed in the 1925 Contest by the winning Army Curtiss R3C-2 (the 1924 Contest had been cancelled, sportingly, by the Americans to allow further development of rival aircraft).

The early 1920s also saw the start of the famous National Air Races, as such, leading to competitions between the latest military pursuit aircraft and specially-built, often controversial racing types. But, although air racing in America continued up to the beginning of the Second World War, interest by the military waned by the 1930s, as its leaders considered that little could be gained from the competitions in the form they had taken.

The first major post-First World War air race to be held in America was the Pulitzer Trophy Race, held at Mitchell Field, Long Island, New York, on Thanksgiving Day 1920. Sponsored by the American newspaper owner Ralph Pulitzer, the competition was open to anyone, the only restriction being that competing aircraft should have landing speeds of less than 75 mph (120·7 km/h). Each lap of the triangular course was 29·02 miles (46·7 km) long, and four laps were flown. The winning aircraft and pilot, together with the winners of subsequent Pulitzer races, are listed below.

The first Air League Challenge Cup race was flown at Croydon on 17 September 1921. Because there were so few civil aircraft then available for competitive events, it was decided to award the cup initially to RAF teams. The first winners were a team from No 24 Squadron, then based at RAF Kenley, Surrey.

The first King's Cup Air Race was held on 8–9 September 1922 from Croydon to Glasgow and back, a distance of 810 miles (1304 km). Originating from a suggestion to hold an annual air race open to British aircraft, and with the object of encouraging sporting flying, HM King George V presented what was originally intended as a Challenge Cup. Subsequently, a new King's Cup was awarded each year. The winner of this first race was Captain F L Barnard, Commodore of the Instone Air Line, flying Sir Samuel Instone's DH4A *City of York* (G-

PULITZER TROPHY RACE WINNERS

Year	Location	Aircraft	Pilot	Average speed
1920	Mitchell Field, Long Island, NY	Verville-Packard VCP-R	Capt Corliss Moseley, USAAS	156·5 mph (251·9 km/h)
1921	Omaha, Nebraska	Navy Curtiss CR-1	Bert Acosta	176·7 mph (284·4 km/h)
1922	Selfridge Field, Michigan	Army Curtiss R-6	1/Lt Russell Maughan, USAAS	205·8 mph (331·3 km/h)
1923	Lambert Field, St Louis, Missouri	Navy Curtiss R2C-1	Lt Alford Williams, USN	243·7 mph (392·15 km/h)
1924	Wright Field, Dayton, Ohio	Verville-Sperry R-3	Lt Harry Mills, USAAS	215·72 mph (347·17 km/h)
1925	Mitchell Field, Long Island, NY	Army Curtiss R3C-1	1/Lt Cyrus Bettis, USAAS	248·97 mph (400·7 km/h)

In 1914 the *Daily Mail* newspaper purchased the prototype Avro 504 which toured the UK giving joyrides. Earlier, the Avro 504 had finished fourth in the 1913 Aerial Derby, only two days after making its first flight, on 18 September

EAMU). With a total flying time of 6 h 32 min 50 s, Captain Barnard recorded an average speed of 123·6 mph (198·9 km/h). Second was Fred Raynham in a Martinsyde F6 (*G-EBDK*) and third Alan Cobham in a de Havilland DH9B (*G-EAAC*).

The first race for the Grosvenor Challenge Cup, presented by Lord Edward Grosvenor for competition by British aircraft with engines not exceeding 150 hp and flown by British pilots, was held at Lympne, Kent, on 23 June 1923. The first winner of the Cup was Flight-Lieutenant W H Longton who, flying the Sopwith Gnu *G-EAGP*, completed the 404-mile (650-km) course at an average speed of 87·6 mph (141 km/h).

The first Schneider Trophy Contest to be won by an American aircraft was that held in September 1923 at Cowes, Isle of Wight, England. The winning aircraft was a Navy Curtiss CR-3 racer, piloted by Lt David Rittenhouse, USN, which achieved a speed of 181·1 mph (291·445 km/h) on its fifth lap. Second place was also taken by an American Navy Curtiss CR-3, piloted by Lt Rutledge Irvine, USN; third place went to Captain Henri Biard of England in a Supermarine Sea Lion III.

The 1924 Schneider Trophy Contest was cancelled. America then won the 1925 Contest, the winner being 1/Lt James Doolittle, USAAS, piloting the Army Curtiss R3C-2 racer. Fastest lap time by Doolittle's R3C-2 was 235·04 mph (378·255 km/h).

The first lightplane competition held in Great Britain was at Lympne, Kent, in October 1923, organised by the Royal Aero Club. The competition was for single-seat light aircraft,

Curtiss R3C-2 flown to victory by Lt "Jimmy" Doolittle in the 1925 Schneider Trophy contest (USAF)

and the prizes included £1000 offered by the *Daily Mail* and £500 offered by the Duke of Sutherland for the longest flight on one gallon of petrol made by an aircraft with an engine not exceeding 750 cc. The prize money for the one-gallon flight was shared between Flight-Lieutenant Walter Longton, who flew an English Electric Wren, and Jimmy H James in an ANEC monoplane, both flying 87·5 miles (140·8 km). Another prize of £500, offered by the Abdulla Co for the highest speed over two laps, was won in a Parnall Pixie by Captain Norman Macmillan, who achieved an average speed of 76·1 mph (122·5 km/h).

The first flight of the first de Havilland Moth prototype G-EBKT was made by Captain Geoffrey de Havilland on Sunday, 22 February 1925, from Stag Lane Aerodrome, Edgware, Middlesex. Sir Sefton Brancker, Director of Civil Aviation, was so impressed with the new aircraft that he recommended the formation of five Government-subsidised flying clubs to be equipped with Moths, and the first such aircraft, G-EBLR, was delivered to the Lancashire Aero Club by Alan Cobham on 21 July 1925. Moths were subsequently manufactured in Australia, Finland, France, America, Canada and Norway, and the type is regarded as being responsible for the initiation of the flying-club movement round the world.

The first Bendix Trophy Race took place in 1931, covering 2043 miles (3288 km) from Burbank, California to Cleveland, Ohio. The winner was Jimmy Doolittle in the Laird Super-Solution biplane, flying at an average speed of 223·038 mph (358·94 km/h).

Scott and Campbell Black arrive at Melbourne

Bendix Trophy Races continued up to the Second World War and were resumed after the war, being extended in 1946 by a special division for jet-powered aircraft. The first winner of the jet division was Col Leon Gray, flying a Lockheed P-80A at an average speed of 494·78 mph (796·27 km/h).

Perhaps the most controversial racing aircraft ever built in America were the Granville brothers' Gee Bee racers of the early 1930s. Conceived to have huge Pratt and Whitney Wasp radial engines mounted in minimum possible airframes, all seven which were built eventually crashed, killing five pilots, but not before one had gained the landplane speed record of 296·287 mph (476·741 km/h) on 3 September 1932, in the hands of Jimmy Doolittle. Gee Bees also won the 1931 and 1932 Thompson Trophy Races.

The first 'trans-World' air race was the MacRobertson Race from England to Australia which started on 20 October 1934. In March 1933 the Governing Director of MacRobertson Confectionery Manufacturers of Melbourne, Sir MacPherson Robertson, offered £15 000 in prize-money for an air race to commemorate the centenary of the foundation of the State of Victoria. (This formed part of a donation of £100 000 placed at the disposal of the Victoria Government in 1933.) The race was won by one of three specially built de Havilland DH88 Comets. A two-seat low-wing monoplane, it was powered by two 230 hp Gipsy Six R engines, each driving a variable-pitch propeller of unusual design. Set in fine-pitch for take-off, they were moved to the coarse-pitch cruise setting by compressed air after the machine was airborne and at suitable height; they could not be recycled back to the fine-pitch setting. Charles W A Scott and Tom Campbell Black were first to cross the finishing line at Flemington Racecourse, Melbourne, in the DH88 *Grosvenor House* (G-ACSS), having completed the 11 333 miles (18 239 km) from Mildenhall, Suffolk, in 70 h 54 min 18 s at an average speed of 158·9 mph (255·7 km/h). Second home in the Handicap Race was, surprisingly, the Douglas DC-2 *Uiver* (PH-AJU) passenger transport aircraft of the Dutch airline KLM, flown by K D Parmentier and J J Moll.

The smallest piloted aeroplane ever flown is the Stits Skybaby biplane, designed and built by Ray Stits at Riverside, Calif, in 1952. It had a wing span of 7 ft 2 in (2·18 m) and a

The world's first all-paper aeroplane (see below) after a flight

length of 9 ft 10 in (3·00 m). Powered by an 85 hp Continental C85 engine, it weighed 452 lb (205 kg) empty and had a top speed of 185 mph (298 km/h).

The first woman in the world to fly faster than the speed of sound was Miss Jacqueline Cochrane, an American cosmetics tycoon, who, flying a North American F-86 Sabre, exceeded the speed of sound on 18 May 1953, and on the same day established a world's speed record for women of 652 mph (1049 km/h).

The lightest aeroplane built and flown in Great Britain is the Ward P46 Gnome, built by Michael Ward of North Scarle, Lincolnshire, and flown on 4 August 1967. Empty it weighs 210 lb (95·3 kg), with a maximum take-off weight of 380 lb (172 kg). Its ceiling, officially imposed at 10 ft (3·05 m) because of the use of materials not covered by air regulations, is probably the **lowest ceiling of any aeroplane!**

The first Transatlantic air race was sponsored by the *Daily Mail* during 1969, for the fastest journey between the top of the Post Office Tower in London and the top of the Empire State Building in New York. The £5000 first prize was won by the Royal Navy entry flown by Lieutenant-Commander Brian Davies, aged thirty-five, and Lieutenant-Commander Peter Goddard, aged thirty-two, in a McDonnell Douglas F-4K Phantom II. Flying time was 4 h 36 min 30·4 s, and the overall time between terminal points was 5 h 11 min 22 s.

The smallest ornithopter (flapping-wing aircraft) in the world is the record-breaking model built by Mr Kenneth B Johnson of Cin-

cinnati, Ohio, with a wingspan of only 5·5 in (14 cm). Another of his model ornithopters, with a wingspan of 1 ft 6 in (0·46 m) and weight of 0·25 oz (7·09 grm), achieved a record flight of 5 min 15·2 s at Akron, Ohio, in 1968.

The world's first all-paper man-carrying aircraft, built entirely of paper, glue and masking tape as an aeronautical teaching aid at Ohio State University, flew on three occasions in August 1970. Towed into the air by a motor car, the all-paper glider had no wheels, but slid over the grass field on a fuselage undersurface of waxed corrugated paper. Maximum airspeed recorded was approximately 60 mph (96 km/h).

Founder member of the Ton-Up Gliding Club was the late Mr James Chapman of Wisbech, Cambridgeshire, who flew in one of the Norfolk Gliding Club's sailplanes on 26 September 1975 at the age of 102. He had made his first-ever flight at the age of 100 in a powered aeroplane, and was 101 when he made his first flight in a hot-air balloon.

The lightest aeroplane yet flown is the Birdman TL-1, the prototype of which first flew on 25 January 1975. It was designed and built by Mr Emmett Tally of Daytona Beach, Florida, and weighed only 100 lb (45 kg) empty. Several examples are now flying.

The fastest 'homebuilt' aircraft ever produced is undoubtedly the Red Baron F-104RB Starfighter, built over a ten-year period by American Darryl Greenamyer. Although similar to Lockheed Starfighters still in military service in various countries, it differs mainly in being built from components acquired from all over the world. It is powered by a General Electric J79-GE-10A turbojet engine of 18000 lb thrust, and is unarmed. Sponsored by the Red Baron Flying Service, Greenamyer attempted to beat the world speed record over a

"Homebuilt" Starfighter; Darryl Greenamyer's F-104RB

Gossamer Condor man-powered aircraft

3 km low-level course in October 1976. Averaging 1010 mph (1625 km/h) in four runs at Mud Lake, Tonopah, Nevada, he bettered the record which is held by a McDonnell Douglas Phantom II at 902 mph (1452 km/h); but a malfunction in the timing equipment meant that Greenamyer's record was not officially recognised by the FAI. Greenamyer is expected to make an attempt at beating the world absolute height record, currently held by the Soviet E-266 (MiG-25).

The first man-powered aircraft to complete a figure-eight flight around two pylons half a mile apart, and thus win the British £50000 Kremer Prize, was the MacCready Gossamer Condor, on 23 August 1977. Average speed during the flight was between 10 and 11 mph (16-17·5 km/h), achieved by the pilot, Bryan Allen, by producing just 0·37 hp. Wing span of the Gossamer Condor was 96 ft 0 in (29·26 m) and all-up weight with the pilot was just 207 lb (94 kg).

AEROBATICS

Aerobatics are almost as old as flying itself, and their development was a measure of the growing confidence in the aeroplane, not only as a flying machine but as a vehicle of sport. The word 'aerobatic' is obviously a contraction of the original 'aerial acrobatics', and as such found its appeal in the dramatic and spectacular, for no sooner had a pilot extricated himself from an unusual attitude in the air than he set about perfecting the manoeuvre in such a manner as to impress his grounded spectators. Of course it was not long before fairly complicated manoeuvres came to be developed for necessity – such as, in air combat, the 'Immelmann turn' – now thought to have been a climbing half-loop with a roll off the top. Yet for many years aerobatics as such remained no more than a spectacle provided by a skilled, individual pilot, performed to display the manoeuvrability of his aeroplane – or simply his own skill and daring. There were certainly few parameters in aerobatics.

More recently aerobatics have been pursued 'scientifically', and since the Second World War national prestige has been upheld through lavish displays of aerobatics performed by formations of interceptor fighters or jet trainers. Almost every major air force in the world now has trained squadrons or flights to provide the spectacle of such displays.

Since 1960 nations have competed in world aerobatic championships, and so complicated and precise have modern aerobatics become in these championship meetings that only a very small handful of specialist aeroplanes, designed and prepared with extreme precision, have been capable – even in the most expert hands – of competing with any chance of success. It is almost certainly true to state that the successful pilots in these championships are the finest exponents of the pure art of flying.

The first aerobatic manoeuvre was undoubtedly the spin. Though nowadays not regarded as an aerobatic but simply as a manoeuvre in which the aeroplane falls in a stalled condition while rolling, pitching and yawing simultaneously, it was originally performed as a spectacular manoeuvre – once the means of recovery had been discovered.

The first British pilot to survive a spin (probably first in the world) was Fred Raynham who, flying an Avro biplane during 1911, stalled while climbing through fog. The stall occurred after he had stooped to adjust his compass as he thought that it was malfunctioning; the next he knew was that he was standing upright on the rudder pedals with his aeroplane whirling round. Quite how he recovered from the spin will never be known, for his recollection was that he *pulled the stick back*; notwithstanding this he caught sight of the ground and was able to perform a controlled landing.

The first pilot to perform, recover from and demonstrate recovery from a spin was Lieutenant Wilfred Parke, RN, on 25 August 1912 on the Avro cabin tractor biplane during the Military Trials of that year. On this occasion Parke and his observer, Lieutenant Le Breton, RFC, were flying at about 600 ft (200 m) and commenced a spiral glide prior to landing; finding that the glide was too steep, Parke pulled the stick back, promptly stalled and entered a spin. With no established procedure in mind for recovery he attempted to extricate himself from danger by pulling the stick further back and applying rudder *into* the direction of spin, and found that the spin merely tightened. After carefully noting this phenomenon he decided, *when only 50 ft (15 m) from the ground* – and from disaster – to reverse the rudder, and the machine recovered instantly. Parke was able to give a carefully reasoned résumé of his corrective actions, thereby contributing immeasurably to the progress of aviation.

The first pilot in the world to perform a loop was Lieutenant Nesterov of the Imperial Russian Army who, flying a Nieuport Type IV monoplane, performed the manoeuvre at Kiev on 27 August 1913.

The first pilot to fly inverted in sustained flight (as distinct from becoming inverted during the course of the looping manoeuvre) was Adolphe Pégoud who, on 21 September 1913, flew a Blériot monoplane inverted at Buc, France. Notwithstanding the above definition, Pégoud's manoeuvre involved two 'halves' of a loop, in that he assumed the inverted position by means of a half-loop, and after sustained inverted flight recovered by means of a 'pull-through'. He thus did not resort to a roll or half-

roll, which manoeuvre had not apparently been achieved at this time. As a means of acclimatising himself for the ordeal of inverted flight, Pégoud had had his Blériot mounted inverted upon trestles and had remained strapped in the cockpit for periods of up to 20 min at a time!

The first woman in the world to experience a loop in an aeroplane was Miss Trehawke Davis, who was taken aloft for the manoeuvre by her protégé, Gustav Hamel (son of an English-naturalised German doctor), at Hendon, probably in September 1913.

The cartwheel manoeuvre (essentially performed in a twin-engined aeroplane) was first demonstrated by Jan Zurakowski in a de Havilland Hornet fighter (two Rolls-Royce Merlin engines) at Boscombe Down, England, in 1945. The same pilot also later gave demonstrations of the manoeuvre in a Gloster Meteor IV jet fighter at the annual SBAC Displays at Farnborough, Hants. It was performed during a vertical full-power climb to the point of near-stall by quickly throttling-back one engine and performing a controlled, vertical wing-over.

The first man to perform an outside loop manoeuvre in America was Jimmy Doolittle in 1927. Two years later he also performed the first totally blind landing.

The 'Derry Turn' was evolved by John Derry, a test pilot of de Havilland Aircraft Co, Ltd, in 1949–50. It was a positive-G turn initiated by rolling in the opposite direction through 270°. It was a fairly spectacular manoeuvre, only ultimately made possible by the availability of sufficient excess engine-power allied with rudder control to keep the nose of the aircraft up at the necessary late stage in the rolling manoeuvre.

The only known occasions on which inverted spins by high-performance jet aircraft have been demonstrated publicly, were the SBAC Displays at Farnborough, Hants, in September 1959 and September 1960. At these displays the Hawker Aircraft Ltd Chief Test Pilot, A W ('Bill') Bedford, flying the demonstration Hunter two-seater *G-APUX*, performed inverted spins of twelve or thirteen turns and used coloured smoke to trace the pattern of his recovery in the sky.

Parachuting has become a popular sport throughout the world since the Second World War. These Spanish parachutists are preparing to jump from a CASA C-212 twin-turboprop light transport

PARACHUTING FOR SPECTACLE, SPORT AND NECESSITY

The first demonstration in the world of a quasi-parachute was given by the Frenchman, Sebastien Lenormand, who in 1783 descended from an observation tower at Montpellier, France, under a braced conical canopy.

The first parachute descent ever performed successfully by man from a vehicle was accomplished by the Frenchman, André Jacques Garnerin, who jumped from a balloon at about 3000 ft (915 m) having ascended from the Parc Monçeau near Paris on 22 October 1797.

The first parachute descent from a balloon in America was that made by Charles Guille who, on 2 August 1819, jumped from a hydrogen balloon at a height of about 8000 ft (2440 m) and landed at New Bushwick, Long Island, N.Y.

The first parachute descent from an aeroplane in America was performed by Captain Albert Berry who, on 1 March 1912, jumped from a Benoist aircraft flown by Anthony Jannus at 1500 ft (460 m) over Jefferson Barracks, St Louis, Mo.

The first parachute descent by a woman from an aeroplane was made by the eighteen-year-old American girl, Georgia ('Tiny') Broadwick who, using an 11 lb (5 kg) silk parachute, jumped from an aircraft flown by Glenn Martin at about 1000 ft (305 m) over Griffith Field, Los Angeles, Calif, on 21 June 1913.

The first parachute drop from an aeroplane over Great Britain was made by W Newell at Hendon on 9 May 1914 from a Grahame-White

The same men, jumping from the port side door of the C-212

Charabanc flown by R H Carr. Newell sat on a short rope attached to the port undercarriage, clutching his 40 lb (18 kg) parachute in his lap; when the aeroplane had climbed to 2000 ft (610 m) F W Gooden, seated on the lower wing, prised Newell off his perch with his foot! The parachute was 26 ft (7·9 m) in diameter and the drop occupied 2 min 22 s.

The first successful use of a free parachute (i.e. a 'free fall') **from an aeroplane** was made by Leslie Leroy Irvin (1895–1965) on 19 April 1919 using a parachute of the pattern he had developed for the US Army. His descent was made from an aircraft flying at 3000 ft (915 m). Irvin's parachute consisted of a body-harness, which had a bag attached to the back containing the canopy and rigging-lines. Pulling a ripcord opened the bag, releasing a spring-ejected pilot parachute which pulled the canopy and rigging-lines from the pack.

The first American to escape from a disabled aeroplane by parachute was Lieutenant Harold R Harris, US Army, who on 20 October 1922 jumped from a Loening monoplane at 2000 ft (610 m) over North Dayton, Ohio.

The greatest altitude from which anyone has ever jumped without a parachute and survived is 22 000 ft (6705 m). In January 1942 Lieutenant (now Lieutenant-Colonel) I M Chisov of the USSR fell from an Ilyushin Il-4 which had been badly damaged. He struck the ground a glancing blow on the edge of a snow-covered ravine and slid to the bottom, sustaining a fractured pelvis and severe spinal damage. (It is estimated that the human body reaches 99 per cent of its low-level terminal velocity after falling 1880 ft (573 m); this is 117–125 mph (188–201 km/h) at normal atmospheric pressure in a random posture, but up to 185 mph (298 km/h) in a head-down position.) **The British record** stands at 18 000 ft (5490 m) set by Flight-Sergeant Nicholas Stephen Alkemade, RAF, who jumped from a blazing Lancaster bomber over Germany on 23 March 1944. His headlong fall was broken by a fir tree, and he landed *without a broken bone* in an 18 in (46 cm) snow-bank.

Squadron Leader Fifield making the first live ejection from an aircraft on the ground

The first use of an ejection seat, to enable a man to escape from an aircraft in flight, occurred on 24 July 1946. This was the date when the first experimental live ejection was made, using a Martin-Baker ejection seat fitted in a Gloster Meteor. With the aircraft travelling at 320 mph (515 km/h), 'guinea pig' Bernard Lynch was shot into the air at a height of 8000 ft (2440 m). In subsequent tests, Lynch made successful ejections at 420 mph (675 km/h) at heights up to 30000 ft (9145 m).

The first man to bail out from an aeroplane flying at supersonic speed and live was George Franklin Smith, aged thirty-one, test pilot for North American Aviation Corporation, who ejected from a North American F-100 Super Sabre on 26 February 1955 off Laguna Beach, Calif. After failure of the controls in a dive. Smith fired his ejector seat at an indicated speed of Mach 1·05 or more than 700 mph (1125 km/h). After being unconscious for five days Smith made an almost complete recovery from his injuries which included haemorrhaged eyeballs, damage to lower intestine and liver, knee-joints and eye retina. Within nine months he was passed fit to resume flying.

The first member of the Royal Air Force to survive a supersonic ejection (and the second man in the world) was Flying Officer Hedley Molland who escaped from a Hawker Hunter fighter on 3 August 1955. Flying at 40000 ft (12190 m). the aircraft went into an uncontrol-

lable dive. All of Flying Officer Molland's attempts to regain control failed, and by the time he ejected his stricken machine was travelling at an estimated Mach 1·10, its height about 10000 ft (3050 m). Descending in the sea he was picked up by a tug, and recovered in hospital from his injuries which included a broken arm

Lieutenant-Colonel William H Rankin (US Marine Corps)

(caused by flailing in the slipstream) and a fractured pelvis.

The world's first live ejection from an aeroplane travelling at speed on the ground was achieved by Squadron Leader J S Fifield, DFC, AFC, on 3 September 1955 at Chalgrove Airfield, Oxfordshire, when he was ejected from the rear cockpit of a modified Gloster Meteor 7 piloted by Captain J E D Scott, Chief Test Pilot of Martin-Baker Ltd, manufacturers of the ejector seat. Speed of the aircraft at the moment of ejection was 120 mph (194 km/h) and the maximum height reached by the seat was 70 ft (21 m) above the runway.

The greatest altitude from which a successful emergency escape from an aeroplane has been reported is 56 000 ft (17 070 m). At this altitude on 9 April 1958 an English Electric Canberra bomber exploded over Monyash, Derbyshire, and the crew, Flight-Lieutenant John de Salis, twenty-nine, and Flying Officer Patrick Lowe, twenty-three, fell free in a temperature of −70°F. (−56·7°C) down to an altitude of 10 000 ft (3050 m), at which height their parachutes were deployed automatically by barometric control.

The longest recorded parachute descent was that by Lieutenant-Colonel William H Rankin of the US Marine Corps who, on 26 July 1959, ejected from his LTV F8U Crusader naval jet fighter at 47 000 ft (14 326 m). Falling through a violent thunderstorm over North Carolina, his descent took 40 min instead of an expected time of 11 min as he was repeatedly forced *upwards* by the storm's vertical air currents.

The greatest altitude from which a man has fallen and the longest delayed drop ever achieved by man was that of Captain Joseph W Kittinger, DFC, aged thirty-two, of the US Air Force, over Tularosa, New Mexico, USA, on 16 August 1960. He stepped out of a balloon gondola at 102 200 ft (31 150 m) for a free fall of 84 700 ft (25 816 m) lasting 4 min 38 s, during which he reached a speed of 614 mph (988 km/h) despite a stabilising drogue and experienced a minimum temperature of −94°F (−70°C). His 28 ft (8·5 m) parachute deployed at 17 500 ft (5334 m) and he landed after a total time of 13 min 45 s. The step by the gondola was inscribed 'This is the highest step in the world.'

The speed of 614 mph (988 km/h) reached by Kittinger during his fall represents a Mach No of 0·93 in the Stratosphere and would have been reached at an altitude of about 60 000 ft (18 300 m); thereafter his fall would have been retarded fairly rapidly to less than 200 mph (322 km/h) as he passed through the Tropopause at about 36 000 ft (11 000 m). The speed of 614 mph (988 km/h) almost certainly represents the greatest speed ever survived by a human body not contained within a powered vehicle beneath the interface (i.e. within the earth's atmosphere).

The longest delayed drop by a woman was 46 250 ft (14 100 m) made by the Russian woman parachutist, O. Komissarova, on 21 September 1965.

The world's 'speed record' for parachute jumping is held by Michael Davis, aged twenty-four, and Richard Bingham, aged twenty-five, who made eighty-one jumps in 8 h 22 min at Columbus, Ohio, USA, on 26 June 1966.

The British record for a delayed drop by a group of parachutists stands at 39 183 ft (11 942 m), achieved by five Royal Air Force parachute jumping instructors over Boscombe Down, Wiltshire, on 16 June 1967. They were Squadron Leader J Thirtle, Flight-Sergeant A K Kidd, and Sergeants L Hicks, P P Keane and K J Teesdale. Their jumping altitude was 41 383 ft (12 613 m).

The greatest number of parachute jumps made by one man is over 5000 by Lieutenant-Colonel Ivan Savkin of the USSR, born in 1913, who reached 5000 on 12 August 1967. It has been calculated that since 1935 Savkin has spent 27 h in free fall, 587 h floating and has dropped 7800 miles (12 550 km). **The largest number of jumps made by a Briton** is believed to be the 1601 descents made by Flight-Lieutenant Charles Agate, AFC (born March 1905), all with packed parachutes, between 1940 and 1946.

The most northerly parachute jump was that by the Canadian, Ray Munro, aged forty-seven, of Lancaster, Ontario, who on 31 March 1969 descended on to the polar ice cap at 87° 30′ N. His eyes were frozen shut instantly in the temperature of −39°F (−39·5°C).

The greatest landing altitude for a parachute jump was 23 405 ft (7134 m), the height of the summit of Lenina Peak on the borders of Tadzhikistan and Kirgiziya in Kazakhstan, USSR. It was reported in May 1969 that ten Russians had parachuted on to this mountain peak but that four had been killed.

APPENDIX I

PROGRESSIVE WORLD ABSOLUTE SPEED RECORDS ACHIEVED BY MAN IN THE ATMOSPHERE

Those entries marked with an asterisk represent accurately measured speeds not ratified as World records by the Fédération Aéronautique Internationale. These speeds were often achieved during one run of several, from which the mean was submitted for ratification as a record; they are included to indicate the progressive highest speed achieved by man.

Speed mph	km/h	Pilot	Nationality	Aircraft	Location of achievement	Date
25·64*	41·27	Alberto Santos-Dumont	Brazil	Santos-Dumont '14bis'	Bagatelle, France	12 Nov 1906
32·72*	52·66	Henry Farman	France	Voisin biplane	Issy-les-Moulineaux, France	26 Oct 1907
34·03	54·77	Paul Tissandier	France	Wright biplane	Pau, France	20 May 1909
43·34	69·75	Glenn Curtiss	USA	Herring-Curtiss biplane	Reims, France	23 Aug 1909
46·17	74·30	Louis Blériot	France	Blériot monoplane	Reims, France	24 Aug 1909
47·84	76·99	Louis Blériot	France	Blériot monoplane	Reims, France	28 Aug 1909
48·20	77·57	Hubert Latham	France	Antoinette monoplane	Nice, France	23 Apr 1910
66·18	106·50	Léon Morane	France	Blériot monoplane	Reims, France	10 July 1910
68·18	109·73	Alfred Leblanc	France	Blériot monoplane	Reims, France	29 Oct 1910
69·46	111·79	Alfred Leblanc	France	Blériot monoplane	Belmont Park, NY	12 Apr 1911
74·40	119·74	Édouard Nieuport	France	Nieuport biplane		11 May 1911
77·67	124·99	Alfred Leblanc	France	Blériot monoplane		12 June 1911
80·80	130·04	Édouard Nieuport	France	Nieuport biplane	Châlons, France	16 June 1911
82·71	133·11	Édouard Nieuport	France	Nieuport biplane	Châlons, France	21 June 1911
90·18	145·13	Jules Védrines	France	Deperdussin monoplane	Pau, France	13 Jan 1912
100·21	161·27	Jules Védrines	France	Deperdussin monoplane	Pau, France	22 Feb 1912
100·99	162·53	Jules Védrines	France	Deperdussin monoplane	Pau, France	29 Feb 1912
103·64	166·79	Jules Védrines	France	Deperdussin monoplane	Pau, France	1 Mar 1912
104·32	167·88	Jules Védrines	France	Deperdussin monoplane	Pau, France	2 Mar 1912
106·10	170·75	Jules Védrines	France	Deperdussin monoplane		13 July 1912
108·16	174·06	Jules Védrines	France	Deperdussin monoplane	Chicago, Illinois, USA	9 Sept 1912
111·72	179·79	Maurice Prévost	France	Deperdussin monoplane		17 June 1913
119·22	191·87	Maurice Prévost	France	Deperdussin monoplane	Reims, France	27 Sept 1913
126·64	203·81	Maurice Prévost	France	Deperdussin monoplane	Reims, France	29 Sept 1913

First World War: No reliable international records

Speed mph	km/h	Pilot	Nationality	Aircraft	Location of achievement	Date
171·01	275·22	Sadi Lecointe	France	Nieuport-Delage 29		7 Feb 1920
176·12	283·43	Jean Casale	France	Blériot monoplane		28 Feb 1920

mph	km/h	Pilot	Country	Aircraft	Place	Date
181·83	292·63	Baron de Romanet	France	Spad biplane		9 Oct 1920
184·51	296·94	Sadi Lecointe	France	Nieuport-Delage 29		10 Oct 1920
187·95	302·48	Sadi Lecointe	France	Nieuport-Delage 29		20 Oct 1920
191·98	308·96	Baron de Romanet	France	Spad biplane		4 Nov 1920
194·49	313·00	Sadi Lecointe	France	Nieuport-Delage 29	Villesauvage, France	12 Dec 1920
210·64*	339·00	Sadi Lecointe	France	Nieuport-Delage 29	Villesauvage, France	25 Dec 1921
205·20	330·23	Sadi Lecointe	France	Nieuport-Delage 29	Villesauvage, France	20 Sept 1922
211·89	341·00	Sadi Lecointe	France	Nieuport-Delage 29		21 Sept 1922
222·93	358·77	Brig-Gen W A Mitchell	USA	Curtiss HS D-12	Detroit, Michigan, USA	13 Oct 1922
243·98*	392·64	Brig-Gen W A Mitchell	USA	Curtiss HS D-12	Detroit, Michigan, USA	18 Oct 1922
233·00	374·95	Sadi Lecointe	France	Nieuport-Delage 29		15 Feb 1923
236·54	380·67	Lt R L Maughan	USA	Curtiss R-6	Mitchell Field, NY, USA	29 Mar 1923
255·40	411·04	Lt A Brown	USA	Curtiss HS D-12	Mitchell Field, NY, USA	2 Nov 1923
267·16	429·96	Lt Alford J Williams	USA	Curtiss R-2 C-1	Mitchell Field, NY, USA	4 Nov 1923
270·50*	435·30	Lt Alford J Williams	USA	Curtiss R-2 C-1	Mitchell Field, NY, USA	4 Nov 1923
274·21*	441·30	Lt A Brown	USA	Curtiss HS D-12		4 Nov 1923
278·47	448·15	Adj Chef A Bonnet	France	Ferbois V-2	Istres, France	11 Dec 1924
284·21*	457·39	Flying Officer S Webster, AFC	GB	Supermarine S5	Venice, Italy	26 Sept 1927
297·83	479·21	Maj Mario de Bernardi	Italy	Macchi M-52	Venice, Italy	4 Nov 1927
313·59*	504·67	Maj Mario de Bernardi	Italy	Macchi M-52	Venice, Italy	6 Nov 1927
318·57	512·69	Maj Mario de Bernardi	Italy	Macchi M-52bis	Venice, Italy	30 Mar 1928
348·59*	561·00	Maj Mario de Bernardi	Italy	Macchi M-52bis	Venice, Italy	30 Mar 1928
>370·00*	>595·40	Flying Off R D Waghorn AFC	GB	Supermarine S6	Solent, England	7 Sept 1929
357·67*	575·62	Flying Off R L R Atcherley	GB	Supermarine S6	Ryde, IoW, England	12 Sept 1929
387·74*	624·00	Sqn Ldr A H Orlebar	GB	Supermarine S6B	Ryde, IoW, England	12 Sept 1929
406·94	654·90	Flt Lt G H Stainforth, AFC	GB	Supermarine S6B	Ryde, IoW, England	13 Sept 1931
415·20*	668·20	Flt Lt G H Stainforth, AFC	GB	Supermarine S6B	Ryde, IoW, England	29 Sept 1931
423·76	681·97	Warrant Officer F Agello	Italy	Macchi-Castoldi 72	Lago di Garda, Italy	10 Apr 1934
430·32*	692·53	Warrant Officer F Agello	Italy	Macchi-Castoldi 72	Lago di Garda, Italy	10 Apr 1934
>434·87*	>699·85	Col Bernascori	Italy	Macchi-Castoldi 72	Desenzano, Italy	18 Apr 1934
440·60	709·07	Lt F Agello	Italy	Macchi-Castoldi 72	Lago di Garda, Italy	23 Oct 1934
441·22*	710·07	Lt F Agello	Italy	Macchi-Castoldi 72	Lago di Garda, Italy	23 Oct 1934
463·82	746·45	Flugkapitän Hans Dieterle	Germany	Heinkel He 100V-8	Oranienburg, Germany	30 Mar 1939
469·17	754·97	Flugkapitän Fritz Wendel	Germany	Messerschmitt Bf 109R	Augsburg, Germany	26 Apr 1939
485·89*	781·97	Flugkapitän Fritz Wendel	Germany	Messerschmitt Bf 109R	Augsburg, Germany	29 Apr 1939

Second World War: No reliable international records

216

Speed mph	km/h	Pilot	Nationality	Aircraft	Location of achievement	Date
602·87*	970·23	Sqn Ldr P Stanbury, DFC	GB	Gloster Meteor F4	Moreton Valence, England	19 Oct 1945
606·25	975·67	Grp Capt H J Wilson, AFC	GB	Gloster Meteor F4	Herne Bay, Kent, England	7 Nov 1945
611·07*	983·42	Grp Capt H J Wilson, AFC	GB	Gloster Meteor F4	Herne Bay, Kent, England	7 Nov 1945
615·65	990·79	Grp Capt E M Donaldson, DSO, AFC	GB	Gloster Meteor F4	Rustington, Sussex, England	7 Sept 1946
623·43*	1003·31	Grp Capt E M Donaldson, DSO, AFC	GB	Gloster Meteor F4	Rustington, Sussex, England	7 Sept 1946
623·61	1003·60	Col Albert Boyd	USA	Lockheed P-80R Shooting Star	Muroc, California, USA	19 June 1947
640·60	1030·95	Cdr T F Caldwell, USN	USA	Douglas D-558 Skystreak	Muroc, California, USA	20 Aug 1947
650·78	1047·33	Maj M E Carl, USMC	USA	Douglas D-558 Skystreak	Muroc, California, USA	25 Aug 1947
670·84	1079·61	Maj R L Johnson USAF	USA	North American F-86A Sabre	Muroc, California, USA	15 Sept 1948
698·35	1123·89	Capt J Slade Nash, USAF	USA	N A F-86D Sabre	Salton Sea, California	19 Nov 1952
699·79*	1126·20	Capt J Slade Nash, USAF	USA	N A F-86D Sabre	Salton Sea, California	19 Nov 1952
715·60	1151·64	Lt-Col W F Barnes, USAF	USA	North American F-86D Sabre	Salton Sea, California	16 July 1953
727·48	1170·76	Sqn Ldr Neville Duke, DSO, OBE, DFC, AFC	GB	Hawker Hunter 3	Littlehampton, Sussex, England	7 Sept 1953
741·50*	1193·33	Sqn Ldr Neville Duke, DSO, OBE, DFC, AFC	GB	Hawker Hunter 3	Littlehampton, Sussex, England	31 Aug 1953)
735·54	1183·74	Lt-Cdr M Lithgow, OBE	GB	Supermarine Swift 4	Libya, Africa	25 Sept 1953
752·78	1211·48	Lt-Cdr J B Verdin, USN	USA	Douglas F4D-1 Skyray	Salton Sea, California	3 Oct 1953
754·99	1215·04	Lt-Col F K Everest, USAF	USA	North American YF-100A Super Sabre	Salton Sea, California	29 Oct 1953
822·09	1323·03	Col H A Hanes, USAF	USA	North American F-100C Super Sabre	Edwards Air Force Base, California, USA	20 Aug 1955
1131·76	1821·39	Lt P Twiss, OBE, DSC	GB	Fairey Delta 2	Chichester, Sussex, Eng	10 Mar 1956
1207·34	1943·03	Maj Adrian Drew, USAF	USA	McDonnell F-101A Voodoo		12 Dec 1957
1403·79	2259·18	Capt W W Irvin, USAF	USA	Lockheed F-104A Starfighter	Southern California, USA	16 May 1958
1483·51	2387·48	Col G Mosolov	USSR	Mikoyan Type E-66	Sidorovo, Tyumenskaya, USSR	31 Oct 1959
1525·93	2455·74	Maj J W Rogers, USAF	USA	Convair F-106A Delta Dart	Edwards Air Force Base, California, USA	15 Dec 1959
1606·51	2585·43	Lt. Col R B Robinson	USA	McDonnell F4H-1F Phantom II	Edwards Air Force Base, California, USA	22 Nov 1961
1665·89	2681·00	Col G Mosolov	USSR	Mikoyan Type E-166	Sidorovo, Tyumenskaya, USSR	7 July 1962

mph	Pilot		Aircraft	Location	Date
2070·10	Col R L Stephens	USA	Lockheed YF-12A	Edwards Air Force Base, California, USA	1 May 1965
2189·00	Capt E W Joersz and Maj G T Morgan Jr	USA	Lockheed SR-71A	Edwards Air Force Base, California, USA	27 July 1976

PROGRESSIVE MAXIMUM SPEEDS ACHIEVED BY THE AMERICAN X-15 ROCKET-POWERED RESEARCH AIRCRAFT

The record speed achieved by Captain Joersz and Major Morgan on 27 July 1976 represents the highest ratified speed record attained by an aeroplane which took off under its own power from the earth's surface. Between 1960 and 1967 however the US Air Force conducted a substantial programme of manned flight trials with the North American X-15 and X-15A-2. Powered by a liquid oxygen and ammonia rocket engine, the X-15 was carried to altitude by a Boeing B-52 before embarking upon ultra-high-speed and high altitude flights, as summarised next.

mph	km/h	Pilot	Date		mph	km/h	Mach No	Pilot	Date
2111	3397	J A Walker	12 May 1960		3647	5869	5·21	R M White	11 Oct 1961
2196	3534	J A Walker	4 Aug 1960		3900	6276	5·74	J A Walker	17 Oct 1961
2275	3661	R M White	7 Feb 1961		4093	6587	6·04	R M White	9 Nov 1961
2905	4675	R M White	7 Mar 1961		4104	6605	5·92	J A Walker	27 June 1962 (X-15A-2)
3074	4947	R M White	21 Apr 1961						
3300	5311	J A Walker	25 May 1961		4250	6840	6·33	W J Knight	18 Nov 1966
3603	5798	R M White	23 June 1961		4534	7297	6·72	W J Knight	3 Oct 1967 (X-15A-2)
3614	5816	J A Walker	12 Sept 1961						
3620	5826	F S Petersen	28 Sept 1961						

PROGRESSIVE WORLD ABSOLUTE HEIGHT RECORDS ACHIEVED BY MAN IN THE ATMOSPHERE

Height ft	m	Pilot	Nationality	Aircraft	Location	Date
508	155	H Latham	GB	Antoinette	Reims, France	29 Aug 1909
984	300	Comte Charles de Lambert	France	Wright	Paris, France	18 Oct 1909
1486	453	H Latham	GB	Antoinette	Châlons, France	1 Dec 1909
3281	1000	H Latham	GB	Antoinette	France	7 Jan 1910
3966	1209	L Paulhan	France	Henry Farman	Los Angeles, USA	12 Jan 1910
4380	1335	W Brookins	USA	Wright	Indianapolis, USA	14 June 1910
4540	1384	H Latham	GB	Antoinette	Reims, France	7 July 1910
6234	1900	W Brookins	USA	Wright	Atlantic City, USA	10 July 1910
6601	2012	A Drexel	USA	Blériot	Lanark, Scotland	11 Aug 1910
8471	2582	Léon Morane	France	Blériot	Deauville, France	3 Sept 1910
8488	2587	G Chavez	France	Blériot	Issy-les-Moulineaux, France	8 Sept 1910

9120	H Wynmalen	France	Henry Farman	Mourmelon, France	1 Oct 1910
9449	A Drexel	USA	Blériot	Philadelphia, USA	Oct 1910
9711	R Johnston	USA	Wright	Belmont Park, USA	31 Oct 1910
10170	G Legagneux	France	Blériot	Pau, USA	8 Dec 1910
10423	M Loridan	France	Henry Farman	Châlons, France	8 July 1911
10466	Capt Félix	France	Blériot	Etampes, France	9 Aug 1911
12828	Roland Garros	France	Blériot XI	St-Malo, France	4 Sept 1911
16076	Roland Garros	France	Blériot XI	Houlgate, France	6 Sept 1912
17880	G Legagneux	France	Morane-Saulnier	Corbeaulieu, France	17 Sept 1912
18405	Roland Garros	France	Morane-Saulnier	Tunis	11 Dec 1912
19291	M Perreyon	France	Blériot XI	Buc, France	11 Mar 1913
20078	G Legagneux	France	Nieuport	St-Raphael, France	28 Dec 1913
33113	Maj R W Schroeder	USA	Lepere	Dayton, USA	27 Feb 1920
34508	Lt J A MacReady	USA	Lepere	Dayton, USA	18 Sept 1921
35242	Sadi Lecointe	France	Nieuport	Villacoublay, France	5 Sept 1923
36565	Sadi Lecointe	France	Nieuport	Issy-les-Moulineaux France	30 Oct 1923
38418	Lt C C Champion	USA	Wright Apache	Washington, USA	25 July 1927
39140	Lt Apollo Soucek	USA	Wright Apache	USA	8 May 1929
41795	W Neuenhofen	Germany	Junkers W34	Dessau	26 May 1929
43166	Lt Apollo Soucek	USA	Wright Apache	Washington, USA	4 June 1930
43976	Capt C F Uwins	GB	Vickers Vespa	Filton, England	16 Sept 1932
44820	G Lemoine	France	Potez 50	Villacoublay, France	28 Sept 1933
47352	Cdr R Donati	Italy	Caproni 161	Rome, Italy	11 Apr 1934
48698	G Détré	France	Potez 50	Villacoublay, France	14 Aug 1936
49944	Sqn Ldr S R Swain	GB	Bristol 138	Farnborough, England	28 Sept 1936
51362	Lt Col M Pezzi	Italy	Caproni 161	Montecelio, Italy	8 May 1937
53937	Flt Lt M J Adam	GB	Bristol 138	Farnborough, England	30 June 1937
56046	Lt Col M Pezzi	Italy	Caproni 161 *bis*	Montecelio, Italy	22 Oct 1938
59445	J Cunningham	GB	de Havilland Vampire 1	Hatfield, England	23 Mar 1948
63668	W F Gibb	GB	English Electric Canberra	England	4 May 1953
65889	W F Gibb	GB	Eng Elec Canberra	England	29 Aug 1955
70308	M Randrup	GB	Eng Elec Canberra	England	28 Aug 1957
76932	Lt Cdr G C Watkins	USA	Grumman F11F-1 Tiger	USA	18 Apr 1958
79452	R Carpentier	France	SO9050 Trident (*F-ZWUM*)	France	2 May 1958
91243	Maj H C Johnson	USA	Lockheed F-104A Starfighter	USA	7 May 1958
94659	Maj V Ilyushin	USSR	Sukhoi T431	USSR	14 July 1959

PROGRESSIVE WORLD ABSOLUTE DISTANCE RECORDS ACHIEVED BY MAN IN THE ATMOSPHERE

Distance miles	km	Pilot	Nationality	Aircraft	Location of start	Date
772 ft	220 m	A Santos-Dumont	Brazil	Santos-Dumont 14bis	Bagatelle, France	12 Nov 1906
2530 ft	771 m	H Farman	France	Voisin	Issy-les-Moulineaux, France	26 Oct 1907
0·62	1	H Farman	France	Voisin	Issy-les-Moulineaux, France	13 Jan 1908
1·25	2·004	H Farman	France	Voisin	Issy-les-Moulineaux, France	21 Mar 1908
2·44	3·925	L Delagrange	France	Voisin	Issy-les-Moulineaux, France	11 Apr 1908
7·92	12·75	L Delagrange	France	Voisin	Centocelle	30 May 1908
14·99	24·125	L Delagrange	France	Voisin	Issy-les-Moulineaux, France	17 Sept 1908
41·38	66·60	Wilbur Wright	USA	Wright	Auvours, France	21 Sept 1908
62	99·8	Wilbur Wright	USA	Wright	Auvours, France	18 Dec 1908
77·48	124·7	Wilbur Wright	USA	Wright	Auvours, France	31 Dec 1908
83·26	134	Louis Paulhan	France	Voisin	Betheny	25 Aug 1909
96·08	154·62	H Latham	GB	Antoinette	Betheny	26 Aug 1909
111·8	180	H Farman	France	Farman	Betheny	27 Aug 1909
145·53	234·21	H Farman	France	Farman	Mourmelon	4 Nov 1909
244	392·75	Jan Olieslagers	Belgium	Blériot	Mourmelon	20 July 1910
289·4	465·72	M Tabuteau	France	Maurice Farman	Etampes, France	28 Oct 1910
320·6	515·9	G Legagneux	France	Blériot	Pau, France	11 Dec 1910
363·35	584·75	M Tabuteau	France	Maurice Farman	Buc, France	30 Dec 1910
388·4	625	Jan Olieslagers	Belgium	Nieuport monoplane	Kiewit	16 July 1911
449·21	722·94	Fourny	France	Maurice Farman	Buc, France	1 Sept 1911
460	740·3	Gobé	France	Nieuport monoplane	Pau, France	24 Dec 1911
628·1	1010·9	Fourny	France	Maurice Farman	Etampes, France	11 Sept 1912
98 556	30 040	Cdr L Flint	USA	McDonnell Douglas F-4 Phantom II	USA	6 Dec 1959
103 389	31 513	Capt J B Jordan	USA	Lockheed F-104C Starfighter	USA	14 Dec 1959
113 891	34 714	Col G Mossolov	USSR	Mikoyan E-66A	USSR	28 Apr 1961
118 898	36 240	A Fedotov	USSR	Mikoyan E-266	USSR	25 July 1973

Distance (km)	Crew	Country	Aircraft	Route/Place	Date
634·5	A Seguin	France	Henry Farman	Buc, France	13 Oct 1913
1967	Capts L Arrachart and H Lemaître	France	Breguet 19	Etampes, France	3–4 Feb 1925
2675	Capt L Arrachart and Adj Arrachart	France	Potez 550	Le Bourget, France	26–27 June 1926
2930	Capt L Girier and Lt Dordilly	France	Breguet 19	Le Bourget, France	14–15 July 1926
3215	Lt Challe and Capt Weiser	France	Breguet 19	Le Bourget, France	31 Aug–1 Sept 1926
3353	Capts D Costes and J Rignot	France	Breguet 19	Le Bourget, France	28–29 Oct 1926
3609·5	Charles Lindbergh	USA	Ryan monoplane	New York to Paris	20–21 May 1927
3911	C D Chamberlin and A Levine	USA	Bellanca	New York, USA	4–6 June 1927
4466·6	A Ferrarin and D Prete	Italy	Savoia-Marchetti S64	Rome, Italy	3–5 July 1928
4912	Capt D Costes and M Bellonte	France	Breguet 19	Le Bourget, France	27–29 Sept 1929
5011	R N Boardman and J Polando	USA	Wright J6	Brooklyn, USA	28–30 July 1931
5309	Sqn Ldr O Gayford and Flt Lt G Nicholetts	GB	Fairey Special monoplane	Cranwell, England	6–8 Feb 1933
5657	Rossi and P Codes	France	Blériot Zapata	New York, USA	5–7 Aug 1933
6306	Col M Gromov, Ing S Daniline and Cmdt A Youmachev	USSR	ANT-25	Moscow to San Jacinto	12–14 July 1937
6658·3	Flt Lt H A V Hogan and Mosson	GB	Vickers Wellesley	Ismailia to Koepang	5–7 Nov 1938
7158·4	Sqn Ldr R Kellett and Flt Lt Gething	GB	Vickers Wellesley	Ismailia to Darwin	5–7 Nov 1938
7158·4	Flt Lt A N Combe and Bornett	GB	Vickers Wellesley	Ismailia to Darwin	5–7 Nov 1938
7916	Col Irving and Lt Col Stawley	USA	Boeing B-29 Superfortress	Northwest to Washington	12 Nov 1945
11235·6	Cdr T Davis and E P Rankin	USA	Lockheed P2V Neptune	Perth	29 Sept–1 Oct 1946
12532·3	Maj Clyde P Evely	USA	Boeing B-52H	Okinawa to Madrid, Spain	10–11 Jan 1962

APPENDIX 2
AVIATION'S WORST DISASTERS

The following table presents details of all known air disasters involving the loss of more than 100 persons' lives.

Date (day/month/year)	Loss of life	Location of Accident	Aircraft involved
18.6.53	129	Tachikawa AFB, Tokyo, Japan; crashed after engine failure on take-off	USAF C-124
30.6.56	128	Grand Canyon, Arizona; two airliners collided.	TWA Super Constellation and United Air Lines DC-7
16.12.60	134	Air collision over Brooklyn, New York.	United Air Lines DC-8 and TWA Super Constellation
4.3.62	111	Near Douala, Cameroun.	Caledonian DC-7C
16.3.62	107	Lost at sea, western Pacific.	Flying Tiger Super Constellation
3.6.62	130	Crashed on take-off at Paris, France.	Air France Boeing 707
22.6.62	113	Guadeloupe, West Indies; crashed in storm.	Air France Boeing 707
3.6.63	101	In sea off British Columbia, Canada.	Chartered DC-7
29.11.63	118	Montreal, Canada; crashed on take-off.	Trans-Canada Airlines DC-8F
20.5.65	121	Cairo Airport, Egypt.	Pakistan Airlines Boeing 720B
24.1.66	117	Mont Blanc, Switzerland.	Air India Boeing 707
4.2.66	133	Tokyo Bay, Japan.	All-Nippon Boeing 727
5.3.66	124	Mount Fuji, Japan.	BOAC Boeing 707
24.12.66	129	Crashed into village in South Vietnam.	Military-chartered CL-44
20.4.67	126	Nicosia, Cyprus.	Chartered Britannia
20.4.68	122	Windhoek, South Africa; crashed on take-off.	South African Airways Boeing 707
16.3.69	155	Maracaibo, Venezuela; crashed after take-off.	Venezuelan Air Lines DC-9
22.12.69	100+	Nha Trang, Vietnam; aircraft overran runway and crashed into school. Many ground casualties.	Air Vietnam DC-6B
16.2.70	102	Into sea off Santo Domingo; engine failure on take-off.	Compania Dominicana DC-9-30
4.7.70	112	Near Barcelona, Spain; crashed into mountain on approach.	Dan-Air Comet 4
5.7.70	108	Toronto, Canada; mis-selection of spoiler, engine caught fire after touch-down.	Air Canada DC-8
30.7.71	162	Morioko, Japan; collision with JASDF F-86F.	All Nippon Boeing 727
4.9.71	111	Juneau, Alaska; crashed into mountain.	Alaska Airlines Boeing 727
7.1.72	104	Ibiza, Balearic Isles; hit hills near airport.	Iberia Caravelle
14.3.72	112	Fujairah; cause unknown.	Sterling Caravelle
5.5.72	115	Palermo, Sicily; crashed into mountain.	Alitalia Douglas DC-8
18.5.72	108	Kharkov, USSR; cause not announced.	Aeroflot Antonov An-10
18.6.72	118	Staines, Middlesex; mis-selection of leading-edge droop, aircraft stalled.	BEA Trident
14.8.72	156	Schönefeld; fire in rear fuselage.	Interflug Il-62
2.10.72	100	Sochi, USSR; cause not announced.	Aeroflot Il-18
13.10.72	176	Moscow, USSR; crashed by outer marker in bad weather.	Aeroflot Il-62

3.12.72	155	Santa Cruz; loss of control on take-off.	Spantax CV-990
29.12.72	100	Everglades, Florida; crashed during turn on overshoot.	Eastern Air Lines TriStar
22.1.73	173	Kano; crashed on landing.	Alia Boeing 707
21.2.73	108	Sinai Desert; shot down by Israeli Phantoms.	Libyan Arab Airlines Boeing 727
10.4.73	105	Basle; attempting to land in snowstorm.	Invicta Vanguard
–.5.73	about 100	Siberia; hijack attempt, resulting in gunfight.	Aeroflot Tu-104
11.7.73	123	Orly, Paris; fire broke out during descent.	Varig Boeing 707
22.12.73	105	Tangier; crashed into high ground.	Sobelair Caravelle
3.3.74	346	Paris, France; explosive decompression caused by cargo door failure.	THY DC-10
22.4.74	106	Bali, Indonesia; impacted volcano at 4000 ft.	Pan American Boeing 707
27.4.74	118	Leningrad, USSR; take-off accident.	Aeroflot Il-18
4.12.74	191	Sri Lanka; impacted high ground during approach.	Martinair DC-8
24.6.75	115	Kennedy Airport, New York; crashed on landing.	Eastern Boeing 727
8.8.75	188	Agadir, Morocco; crashed into a mountain on approach.	Alia Boeing 707
20.8.75	126	Damascus; crashed during landing.	CSA Ilyushin Il-62
5.3.76	120	Yerevan, USSR; malfunction of pressurisation system during approach.	Aeroflot Ilyushin Il-18
10.9.76	176	Zagreb, Yugoslavia; mid-air collision.	British Airways Trident and Inex-Adria DC-9
20.9.76	155	Karatepe Mountains, Turkey; struck mountains.	THY Boeing 727
27.3.77	579	Santa Cruz Airport, Tenerife. Collision on runway.	KLM and Pan American Boeing 747s

During 1975 and 1976, no fewer than 202 passengers and crew of airline flights were killed by terrorist activity or groundbased arms fire which struck the airliners.

BIBLIOGRAPHY

Air Ministry (1948). *The Rise and Fall of the German Air Force, 1933–1945*; Air Ministry, London.

Authors, Various (1965–7). *Aircraft Profiles* (Seven Volumes, 204 aircraft described); Profile Publications, Leatherhead, England.

Babington-Smith, Constance (1961). *Testing Time: The Story of British Test Pilots and their Aircraft*; Harper, New York.

Barnes, C H (1970). *Bristol Aircraft Since 1910*; Putnam, London.

Barnes, C H (1967). *Shorts Aircraft Since 1900*; Putnam, London.

Bergman, Jules (1960). *Ninety Seconds to Space: The X-15 Story*; Doubleday, New York.

Bowers, Peter M (1965). *Boeing Aircraft Since 1916*; Putnam, London.

Brabazon of Tara, Lord (1956). *The Brabazon Story*; Heinemann, London.

Brett, R D (1934). *The History of British Aviation, 1908–1914*; Hamilton, London.

Brewer, Griffith (1940). *Ballooning and its Application to Kite Balloons*; The Air League of the British Empire, London (11th Edition).

Broke-Smith, Brigadier P W L (1968). *The History of Early British Military Aeronautics*; L A Academic Reprint, Chivers, Bath, England.

Bruce, J M (1957). *British Aeroplanes, 1914–1918*; Putnam, London.

Cornish III, Joseph Jenkins (1963). *The Air Arm of the Confederacy*; Richmond Civil War Centennial Committee, Richmond, Virginia, USA.

Davies, R E G (1964). *The History of the World's Airlines*; Oxford University Press, London.

Dixon, C (1930). *Parachuting*; Sampson Low, London.

Dollfus, Charles (1961). *Balloons*; Prentice-Hall International (Transl. Mason), London.

Dollfus, Charles; Beaubois, Henry; Rougeron, Camille (1965). *L'homme, l'air et l'espace*; Les Editions de l'Illustration, Paris.

Dorian, A F and Osenton, James (eds.) (1964). *Dictionary of Aeronautics* (six languages); Elsevier, London.

Feeney, William D (1963). *In Their Honour*; Duell, Sloan and Pearce, New York.

Francillon, René J (1970). *Japanese Aircraft of the Pacific War*; Putnam, London.

Fricker, John and Green, William (1958). *The Air Forces of the World*; Macdonald, London.

Garber, Paul E (1956). *The National Aeronautical Collections*; The Smithsonian Institution, Washington.

Gibbs-Smith, Charles (1970). *Aviation*; Her Majesty's Stationery Office, London.

Goldberg, Alfred (1957). *A History of the United States Air Force, 1907–1957*; D. Van Nostrand Co Inc, New Jersey.

Haddow, G W and Grosz, Peter M (1969). *The German Giants*; Putnam, London.

Hodgson, J E (1924). *The History of Aeronautics in Great Britain*; Oxford University Press, London.

Hoeppner, Von (1920). *Deutschlands Krieg in der Luft*; Berlin.

Hurren, B J (1951). *Fellowship of the Air 1901–1951*; Iliffe and Sons Ltd, London.

Ingells, Douglas J (1966). *The Plane that changed the World*; Aero Publishers Inc, California.

Inoguchi, Rikihei; Nakajima, Tadashi and Pineau, Roger (1959). *The Divine Wind*; Hutchinson, London.

Jackson, A J (1974). *British Civil Aircraft Since 1919* (three volumes); Putnam, London.

Jackson, A J (1962). *De Havilland Aircraft Since 1915*; Putnam, London, 1962.

Jackson, A J (1965). *Avro Aircraft Since 1908*; Putnam, London.

Jackson, A J (1968). *Blackburn Aircraft Since 1909*; Putnam, London.

Jones, H A and Raleigh, Sir Walter (1922–37). *The War in the Air* (six volumes); Oxford University Press, London.

Josephy, Alvin M (Jr) *The Adventure of Man's Flight*; Putnam, London.

Kerr, Mark (1927). *Land, Sea and Air*; Longmans, Green, London.

King, H F *Aeromarine Origins*; Putnam, London.

Lavenere-Wanderley, Lieutenant-General Nelson Freire (Brazilian Air Force Retired). Historia da Força Aérea Brasileira.

Lewis, Peter (1962). *British Aircraft, 1809–1914*; Putnam, London.

Lewis, Peter. *British Racing and Record-Breaking Aircraft*; Putnam, London.

Marsh, W Lockwood (1924). *Aeronautical Prints and Drawings*; Halton and Truscott Smith, London.

Mason, Francis K (1971). *Hawker Aircraft Since 1920*; Putnam, London.

Mason, Francis K and Windrow, Martin C (1969). *Battle Over Britain*; McWhirter Twins, London.

Mason, H M (1967). *The Lafayette Escadrille*; Random House, New York.

McDonough, Kenneth (1966). *Atlantic Wings 1919–1939*; Model Aeronautical Press Ltd, Herts.

McWhirter, Norris and Ross (1955–77). *The Guinness Book of Records* (and US Edition, *The Guinness Book of World Records*) Editions; Guinness Superlatives Ltd, London.

Mondey, David (1975). *The Schneider Trophy*; Robert Hale, London.

Moyes, Philip (1964). *Bomber Squadrons of the RAF*; Macdonald, London.

Neumann (1920). *Die Deutschen Luftstreitkräfte in Weltkriege*; Berlin.

Nowarra, H J and Brown, K S (1958). *Von Richthofen and the Flying Circus*; Harleyford Publications Ltd, Herts.

Obermaier, Ernst (1966). *Die Ritterkreuzträger der Luftwaffe: Jagdflieger, 1939–45*; Verlag Dieter Hoffmann, Mainz, Germany.

Peaslee, B J (1964). *Heritage of Valor: The Eighth Air Force in the Second World War*; Lippincott, New York.

Penrose, Harald (1967). *British Aviation: The Pioneer Years*; Putnam, London.

Penrose, Harald (1969). *British Aviation: The Great War and Armistice*; Putnam, London.

Phelan, Joseph A (1968). *Heroes and Aeroplanes of the Great War, 1914–1918*; Barker, London.

Price, Alfred (1977). *Instruments of Darkness*; Macdonald and Jane's, London.

Price, Alfred (1969). *German Bombers of the Second World War* (two volumes); Lacy, Windsor.

Priller, Josef (1956). *Geschichte eines Jagdgeschwaders: Das JG 26 'Schlageter' 1937–1945*; Kurt Vowinckel Verlag, Heidelberg, Germany.

Ramsden, J M (1976). *The Safe Airline*; Macdonald and Jane's, London.

Rawlings, J (1969). *Fighter Squadrons of the RAF*; Macdonald, London.

Richards, Denis and Saunders, H St J (1953–6). *Royal Air Force 1939–1945* (three volumes); HMSO, London.

Richthofen, General Baron von (1941–4). *Personal Diary*; Karlsruhe Collection, Hamburg.

Robertson, Bruce and others (1959) *Air Aces of the 1914–18 War*; Harleyford Publications Ltd, Herts.

Robertson, Bruce (1971). *British Military Aircraft Serials 1911–1971*; Ian Allan, Shepperton.

Robinson, Douglas H (1971). *The Zeppelin in Combat*; G T Foulis and Co Ltd, Oxfordshire.

Rolt, L T C (1966). *The Aeronauts: A History of ballooning, 1783–1903*; Longmans, Green, London.

Seemen, Gerhard von (1955 et seq). *Die Ritterkreuzträger, 1939–45*; Podzun-Verlag, Bad Nauheim, Germany.

Sims, Edward H (1958). *American Aces of the Second World War*; Macdonald, London.

Stewart, O (1957). *First Flights*; Routledge and Kegan Paul, London.

Stewart, O (1964). *Of Flights and Flyers*; Newnes, London.

Stroud, John (1962). *Annals of British and Commonwealth Air Transport, 1919–1960*; Putnam, London.

Stroud, John (1966). *European Transport Aircraft Since 1910*; Putnam, London.

Stroud, John (1968). *Soviet Transport Aircraft Since 1945*; Putnam, London.

Swanborough, Gordon and Taylor, J W R (1970). *British Civil Aircraft Register*; Ian Allan, Shepperton.

Swanborough, F G and Bowers, P M (1972). *United States Military Aircraft Since 1908*; Putnam, London.

Swanborough, F G and Bowers, P M (1968). *United States Navy Aircraft Since 1911*; Putnam London.

Taylor, H A (1970). *Airspeed Aircraft Since 1931*; Putnam, London.

Taylor, J W R (1974). *Aircraft Aircraft*; Hamlyn, London.

Taylor, J W R (1955). *A Picture History of Flight*; Hulton Press, London.

Taylor, J W R; Jane, Fred T; Grey, C G; Bridgman, Leonard (ed) *Jane's All the World's Aircraft* (various editions, 1909–77); Sampson Low, *et al.*

Taylor, J W R (1958). *CFS*; Putnam, London.

Taylor, J W R and Swanborough, Gordon (1977). *Civil Aircraft of the World*; Ian Allan, Shepperton.

Taylor, J W R and Munson, Kenneth (1971–77). *History of Aviation*; New English Library.

Taylor, J W R and Allward, M F (1965). *Westland 50*; Ian Allan, London.

Taylor, M J H and J W R (1976). *Helicopters of the World*; Ian Allan, Shepperton.

Taylor, M J H (1976). *Jane's Pocket Book of Research and Experimental Aircraft*; Macdonald and Jane's.

Taylor, M J H, J W R and Munson, Kenneth (1975). *Jane's Pocket Book of Major Combat Aircraft*; Macdonald and Jane's.

Taylor, M J H, J W R and Munson, Kenneth (1975). *Jane's Pocket Book of Military Transport and Training Aircraft*; Macdonald and Jane's.

Thetford, Owen J (1971 et seq). *Aircraft of the RAF Since 1918*; Putnam, London.

Thetford, Owen J (1971 et seq). *British Naval Aircraft Since 1912*; Putnam, London.

Thetford, Owen J and Gray, Peter (1962). *German Aircraft of the First World War*; Putnam, London.

Wallace, Graham (1958). *Flying Witness: Harry Harper and the Golden Age of Aviation*; Putnam, London.

Windrow, Martin C (1968–9). *German Fighters of the Second World War* (Two Volumes); Lacy, Windsor, England.

Ziegler, Mano (1976). *Rocket Fighter*; Arms & Armour Press, London.

ACKNOWLEDGEMENTS

The compilers wish to extend their thanks to the many organisations and individuals who gave generous assistance during the preparation of this book, and particularly to those listed below in alphabetical order:

Aero Spacelines Inc.
Chaz Bowyer
British Aircraft Corporation (all Divisions)
J. M. Bruce, M.A., F.R.Hist.S.
Avions Marcel Dassault – Breguet Aviation
General Dynamics Corporation
Charles H. Gibbs-Smith
Hawker Siddeley Aviation Ltd. (all Divisions)
Imperial War Museum
Italian Air Ministry
Ministry of Defence (R.A.F.)
Ministry of Defence (R.N.)
P. J. R. Moyes

N.A.S.A.
Novosti Press Agency
Alfred Price
Rockwell International
Rolls-Royce Ltd.
The Smithsonian Institution
Society of British Aerospace Companies Ltd.
Tass
United States Air Force
United States Marine Corps
United States Navy
Gordon S. Williams

INDEX